Roadmap on
Photonic Crystals

Authors

Kanna Aoki
RIKEN

Toshihiko Baba
Yokohama National University

Takemi Hasegawa
Sumitomo Electric Industries, Ltd.

Hideo Kosaka
NEC Corp.

Masanori Koshiba
Hokkaido University

Hideki Masuda
Tokyo Metropolitan University

Hiroaki Misawa
The University of Tokushima

Hideki T. Miyazaki
National Institute for Materials Science

Susumu Noda
Kyoto University

Masaya Notomi
NTT Basic Research Laboratories

Takashi Sato
Tohoku University

Mitsuo Wada Takeda
Shinshu University

Roadmap on
Photonic Crystals

edited by

Susumu Noda

Kyoto University
Kyoto, Japan

Toshihiko Baba

Yokohama National University
Yokohama, Japan

Published in cooperation with
**Optoelectronic Industry and Technology Development Association
(OITDA)**

Kluwer Academic Publishers

Dordrecht / Boston / London

ISBN 978-1-4419-5357-5

Distributors for North, Central and South America:
Kluwer Academic Publishers
101 Philip Drive
Assinippi Park
Norwell, Massachusetts 02061 USA
Telephone (781) 871-6600
Fax (781) 871-6528
E-Mail <kluwer@wkap.com>

Distributors for all other countries:
Kluwer Academic Publishers Group
Post Office Box 322
3300 AH Dordrecht, THE NETHERLANDS
Telephone 31 78 6576 000
Fax 31 78 6576 474
E-Mail <orderdept@wkap.nl>

 Electronic Services <http://www.wkap.nl>

Library of Congress Cataloging-in-Publication

Roadmap on photonic crystals/edited by susumu Noda, Toshihiko Baba.
   p.cm.
   "Published in cooperation with Optoelectronic Industry and Technology Development
Association (OITDA)."
   Includes bibliographical references and index.

   1.   Photons. 2. Crystal optics. I. Noda, Susumu. II. Baba, Toshihiko.
   III. Optoelectronic Industry and Technology Development Association (Japan)

QC793.5.P42R63 2003
548'.9—dc21

                                                    2003047502

CONTENTS

PREFACE

Photonic crystals are a new type of optical material in which the refractive index changes periodically. They are likely to provide an exciting new tool for the manipulation of photons and has received keen interest from academia and industry. The book *Roadmap on Photonic Crystals* was first written in Japanese and published in 2000. The original book, written by 15 specialists summarizes the various technological aspects such as the background of photonic crystals, basic theories, numerical simulations, crystal structures, fabrication processes, evaluation methods, and proposed applications, and included a roadmap addressing future development and applications. The main purpose of the book was to help promote the development of commercially available applications of photonic crystals in industry. It was based on proposals put forth by a Japanese association of optics (Optoelectronic Industry and Technology Development Association, OITDA) as a project called *Breakthrough Forum in Optics in Japan*. The book has been well received amongst Japanese industrial scientists, and over 800 copies have been distributed to date. We hope it has benefited the industry figures by promoting the understanding of photonic crystals and its applications.

Two years have now passed since the publication of the first edition and the content of the book deserved revision in light of the fast pace of developments in this field. Fortuitously Drs. Michael S. Hackett and Ana Bozicevic, Editors at Kluwer Academic Publishers, had been considering a book on photonic crystals and sought to undertake an English translation, bringing it forward to a wider audience. The latest edition provides summaries of critical aspects of photonic crystals, and it is hoped that it will be useful to both graduate students and post-doctoral researchers at universities, as well as industrial scientists.

We sincerely thank Dr. Satoshi Ishihara (OITDA) and Dr. Teruo Sakurai (Femtosecond Technology Association), who originally promoted this book and continually supported the present undertaking. We are also grateful to Dr. Yuich Ono (OITDA), who worked tirelessly to arrange the original Japanese version. We greatly appreciate the skillful engagement of Dr. Kanna Aoki of RIKEN, who devoted much of her time to this demanding translation project. Thanks are also given to CREST (Core Research for Evolutional Science and Technology) for the continuous support on this project.

Due to the rapid advances occurring in the photonic crystals research, the present book may not encompass all facets of the area. Nonetheless, we hope that it proves to be helpful by advancing knowledge in the field, and that criticism from readers will find its way into an extended edition in the near future.

Susumu Noda
Kyoto University
Kyoto, Japan

Toshihiko Baba
Yokohama National University
Yokohama, Japan

December, 2002

Chapter 1

PHYSICAL AND EXPERIMENTAL BACKGROUND OF PHOTONIC CRYSTALS

1.1 INTRODUCTION

Nearly fifteen years have passed since the study of *photonic crystals* first commenced. Initially, this study was undertaken as it was seen as an interesting new area in physics. In the last five years, an increasing number of studies have been carried out on device applications in applied physics and engineering in addition to fundamental studies since a breakthrough on the realization of photonic crystals in optical regime has been achieved. The unique properties of photonic crystals have also led to their studies being recognized as a new and major field in optoelectronics. Moreover, study of the physics of photonic crystals continues to grow, drawing on many other scientific fields such as radio techniques, chemistry, precision machinery, acoustics, and so on. This chapter chronologically introduces the milestones reached in the study and development of photonic crystals, as well as some of their physical background.

1.2 INITIAL DEVELOPMENTS OF THE PHOTONIC CRYSTAL RESEARCH

The progress made in photonic crystal research cannot be separated from the evolution of the concept of the *photonic band*. This term describes the optical dispersion relation in a photonic crystal, between a given frequency and the corresponding wavevector, which is generated in analogy to the electron energy band that usually appears in textbooks of solid state physics. That is to say, the relation between electron energy E and momentum p corresponds to the relation between an optical frequency ω and wavevector k. The origin of this correspondence is from the comparison between the quantum mechanical Schrödinger equation for electrons and the electromagnetic wave equation for light.

The term *band of photons* was first used in optics by Otaka (Univ. of Tokyo, though presently at Chiba Univ.) in 1979.[1] At that time it was developed as one of the theories used to explain Bragg diffraction, so its application to the concept of photonic crystals was still not yet clear. In 1987, Yablonovitch (Bellcore, presently at UCLA) proposed the application of crystal engineering technology (for example, *photonic atoms, unit cells, reciprocal spaces*, and *Brillouin zones*) to photonic crystal development. He proposed the possibility of controlling the spontaneous emission of photons within the *photonic bandgap (PBG)*. This is the prohibited band, meaning that propagation of light in this photon energy range is inhibited in all directions in such a material.[2] At almost the same time, John (Princeton Univ. presently at Toronto Univ.) argued that the phenomenon of light localization occurs in a uniform photonic crystal, when randomness is introduced in photonic atom arrangement. This localization would be due to a similar phenomenon to that known as *Anderson localization* in electronic systems.[3] Yablonovitch et al. described a localization phenomenon in more microscopic disorder as *doping*.[4] In 1946, Purcell (Harvard Univ.) had predicted that in a microcavity, the spontaneous emission rate will change in proportion with the light localization and cavity Q-value.[5] Consequently, it has been postulated that fast and strong spontaneous emission would occur in a photonic crystal with impurities or artificial defects under these conditions of strong localization and high cavity Q-values.

It is well known that as the angle of incidence of a light beam changes, the reflection band (stopband) also shifts until at last it disappears in a one-dimensional (1D) periodic structure, like a multilayer film or diffraction grating. The PBG can be explained by a structure having a common stopband for all incident angles of light. Early researchers investigated structures that generate a PBG, after establishing techniques for calculating the photonic band, which were based on ideas taken from solid state crystal engineering theory. Leung et al. (Politech. Univ.) and other groups determined the eigenvalue equation by applying an periodic boundary condition on the scalar approximated wave equation, and then proceeded to calculate many photonic band curves. This is the so-called *plane wave expansion* (PWE) method. They demonstrated a wide PBG in a face-centered-cubic (fcc) lattice, whose first Brillouin zone is the closest to a sphere that has the best isotropic spectral characteristics for all incident angles. In 1990, more rigorous formulation was developed, which took into account the vector characteristics of light. Then the incomplete formation of a PBG was discovered. This was for the case of symmetrical photonic atoms in a fcc lattice, and it was found to be due to the degeneracy of polarized light.[6] Following this discovery, Ho, Soukoulis et al. (Iowa State Univ.) found that the combination of spherical photonic atoms in a diamond structure generated a complete PBG.[7] Moreover, Yablonovitch et al. reported that a PBG can even be formed in fcc lattices, by drilling dielectrics to form a structure named as *Yablonovite*, which has three-fold symmetry.[8] They also made it possible to calculate the impurity states by means of a band calculation method known as the *supercell method*.[4] Pendry et al. (Imperial Collage London Univ.) developed a transfer matrix method, which is a combination of finite element analysis and multilayer analysis methods.[9] This made it possible to analyze transmission characteristics effectively for finite periodic structures.

In parallel with these theoretical studies, progress was also made with the experimental challenges of fabrication. At that time, it was difficult to fabricate fine periodic structures capable of operating at optical frequencies. Yablonovitch et al. made a large dielectric structure with a periodicity in the millimeter range, by means of mechanical techniques. These structures were irradiated with micro- to millimeter-waves from various angles. It was shown that the frequency range of decreasing transmission coincided with the theoretical PBG, and that transmission peaks corresponding to the impurity levels introduced into the crystal were also present.[4] Most of the above achievements are reported in Ref. 10.

1.3 PROGRESS IN PHOTONIC CRYSTAL RESEARCH

The next stage in the evolution of photonic crystal studies was the period leading to the term *photonic crystal* becoming widely accepted. This stage can be characterized by three key points: (i) the discovery of new PBG structures and progress in the speed and flexibility of calculations, (ii) the challenges of developing crystals that operate in the optical frequency regime, and (iii) the study of photonic crystals made of metals, which could operate in the microwave range.

On point (i), much progress was made in the design of photonic crystals and calculations of their theoretical band structure. PBGs were calculated for many three-dimensional (3D) structures based on the diamond structure and the fcc structure composed of asymmetric photonic atom, including the *woodpile* or *stacked-stripe* structure proposed and investigated by Ho, Soukoulis, Ozbay (Iowa State Univ., presently at Bilkent Univ.) et al.[11,12] Also, two-dimensional (2D) photonic crystals were investigated. When light travels parallel to the 2D region, the modes can be decomposed into two separate polarizations, one perpendicular and the other parallel to the 2D plane, and the band structure of each mode can be discussed. Joannopoulos et al. (MIT) showed the existence of the so-called 2D PBG, in which the

PBGs of these two modes overlap each other. The crystal structure designed to achieve the 2D PBG was an array of cylindrical holes in a dielectric material, arranged in a triangular and honeycomb structures. They reported their work on photonic band calculations, also including work on this 2D PBG material, in 1995,[13] their book being a good source of material for learning the basic theory of photonic crystals. Sakoda (Hokkaido Univ., presently at Nat. Inst. Mat. Sci.) developed a PWE technique that is effective for analyzing optical transmission characteristics in finite periodic structures. He classified the photonic crystal modes into two types, ones which are able to couple to the extrinsic optical field, and ones which do not couple.[14] Joannopoulos et al. applied the finite difference time domain (FDTD) method, which was regularly used in the field of electromagnetic waves, to photonic crystals. They then showed that line defects in 2D crystals function as a waveguide for light. The most technologically important discovery in this regard was that propagation with low loss through such photonic crystal waveguides was possible, even when they bend at right angles.[15] This has triggered recent studies into the application of photonic crystals to photonic integrated circuits.[16] Even though the FDTD method has a high degree of flexibility, it is computationally intensive, requiring large computer memories and long calculation time. On the other hand, the scattering matrix method developed by Taybe et al. (Saint Gerome Sci. Tech. Univ.) made it possible to analyze light propagation in a finite periodic crystal, including defects with an arbitrary position and size. Despite the constraints for structures of the photonic atom, this computation can be made in much shorter time.[17]

On point (ii), the first steps towards creating 3D photonic crystals working in the optical frequency range were made by Scherer et al. (Caltech) and Noda et al. (Kyoto Univ.). Scherer et al. attempted to evaluate a Yablonovite structure with submicron periodicity in the near-IR range, which was fabricated on a GaAs using a dry etching technique.[18] Noda et al. proposed a wafer-fusion technique to realize the stacked-stripe structure,[19] which can make it possible to produce a PBG and introduce defects and/or light-emitters.

On the other hand, Gurley et al.[20] (Sandia Nat. Lab.) and Krauss, De La Rue et al.[20] (Univ. Glasgow, Krauss presently being at Univ. St. Andrews) fabricated 2D crystal structures with a high aspect ratio on the GaAs based semiconductors, using a dry etching technique.[21] Krauss et al. confirmed a PBG in the 2D plane using optical I/O waveguides.[22] Baba et al. (Yokohama Nat. Univ.) fabricated 2D crystals using InP based semiconductors, and observed the light emission characteristics.[23] Inoue et al. (Hokkaido. Univ.) reported a 2D silica photonic crystal fabricated from a fiber plate.[24] Grüning et al. (Siemens) demonstrated a 2D silicon photonic crystal[25] having holes with a very high aspect ratio, created by anodization techniques.[26] However, at that time, the study of 2D crystals was primarily focused on increasing the etching aspect ratio sufficiently, so differed from the direction of today's research into slab-type 2D structures.

On point (iii), studies of metals and their relevance to photonic crystal development began when the photonic band was re-calculated by Ho et al.,[27] Pendry et al.,[28] and so on to include the frequency dependence of the metal dielectric constant. This demonstrated that the generation of wide PBGs for electromagnetic waves was possible. A wide PBG was shown to be possible by use of a metal cube arrangement buried in a dielectric material, by Brown et al. (MIT),[29] and GHz frequency PBGs could be generated using a wire mesh of metal, as shown by Yablonovitch et al.[30]

The stage of development, as described above, represents a study of applied physics and engineering, including device and processing techniques, related to the development of the pure physics topic of photonic crystals. However, due to the then perceived lack of applications and realizations, only a limited number of specialists attended the various societies and conferences on optoelectronics to discuss this new expanded field.

1.4 IN THE SPOTLIGHT

Progress in this area of research from 1998 to the present has been remarkable. Indeed, the scientific and industrial community began to pay more attention to photonic crystals thanks to these recent achievements, the most notable of which are outlined below.

The development of 3D crystals with complete PBGs has progressed remarkably. Nanotechnology has been applied to the construction of artificial crystals for the realization of 3D crystals with a complete PBG in the optical regime. Noda et al. made progress in this area and succeeded in the demonstration of a complete PBG with an effect as large as 40 dB seen in the optical communication wavelength range.[31,32] They also designed and fabricated 3D waveguides[33] with a right angle bend, made by introducing defects into the 3D crystal. At the same time, Lin et al. (Sandia Nat. Lab.) used silicon micro-machining techniques for the realization of large scale 3D crystals, and successfully demonstrated a complete PBG in the similar wavelength range.[34] Other 3D PBG structures were also proposed and tried to fabricate, i.e. a spiral crystal by Noda et al.[35] and John et al.[36], and a crystal composed of multilayers and holes by Joannopoulos et al.[37] and Notomi et al. (NTT).[38]

There are some other approaches to the fabrication of large and uniform 3D crystals, although with most of them it is difficult to generate PBGs as they are. The self-organization of nano-spheres has been studied by many groups for a long time in order to construct *opal crystals*. Torres et al. (Univ. Wuppertal),[39] Meseguer et al. (Univ. Polytech. Valencia),[40] Yoshino et al. (Osaka Univ.),[41] Vos et al. (Amsterdam Univ.),[42] Norris et al. (NEC, presently at Univ. Minnesota)[43] and many other groups experimentally demonstrated such opal crystals and two additional techniques; one is the infiltration of a high refractive index material in opal crystals to invert the index profile (inverse opal). Miyazaki et al. (Univ. Tokyo, presently at Nat. Inst. Mat. Sci.) demonstrated a one-by-one manipulation technique of nano-spheres and other nano-elements to construct an arbitrary structure of photonic crystal.[44] Kawakami et al. (Tohoku Univ.) established a fabrication technique using bias sputtering of multilayer on a patterned substrate (autocloning technique).[45] Kawata et al. (Osaka Univ.),[46] Misawa et al. (Tokushima Univ.),[47] Turberfield et al. (Oxford Univ.)[48] and some other groups reported the fabrication of polymer photonic crystals utilizing an interference pattern exposure and scanning exposure of a focused light beam. The spatial resolution of the latter technique was improved by using multi-photon absorption phenomena.

Important progress was also made in 2D crystal fabrication. An aspect ratio of at least 20 is required in a deep crystal structure in order for a PBG effect to be seen that fits the calculated results of the 2D plane, in which light propagates forward expanding into both upper and lower space. As it is not trivial to realize this aspect ratio for a sub-micron periodicity, researchers continue to investigate and extend the various competing manufacturing technologies: dry etching, selective growth, anodic oxidation, and so on. However, a structure known as a *photonic crystal slab* has the potential to drastically reduce the required aspect ratio. This structure is composed of a thin film of semiconductor having an arrangement of holes, which is sandwiched between materials with a low refractive index. This structure is utilized to create a strong optical confinement effect in the upper and lower regions, due to a large refractive index contrast. Using this structure, Scherer et al.[49] and Lee et al (KAIST)[50] achieved point defect laser oscillation by photopumping under low temperature conditions. At present, low threshold room temperature operation has been achieved in various modified structures.

Noda et al. proved that by current injection, the single longitudinal and lateral mode operation of a wide area photonic crystal laser at room temperature was possible. To demonstrate this, they used a 2D photonic crystal in which the 2D photonic crystal slab and an active material were joined using wafer-fusion techniques.[51,52] The oscillation principles of this laser is based on 2D band-edge effect, where the standing wave is formed, and not based on the point defect laser. Meier et al. (Lucent Tech.) also reported the 2D photonic crystal laser based on the band-edge effect by using organic system operated by photopumping.[53] Not only lasers but also light emitting diodes (LEDs) were studied as an application of 2D photonic crystals. Joannopoulos et al. described the *lightcone* in the band diagram of the photonic crystal slab, meaning the leaky condition of in-plane modes. They theoretically predicted a high light extraction efficiency in photonic crystal slab LEDs.[54] This prediction was verified to be correct in some later experiments, so the idea is now at the stage of being considered for further development into a commercial device.

Baba et al. demonstrated light propagation in a line defect waveguide with 60° bends using a photonic crystal slab on a SiO₂/InP substrate.[55] The research on such a waveguide was done worldwide as it could be a key component for a photonic crystal integrated circuit. Noda et al. pointed out that the theoretical loss-less condition of this type of waveguide is obtained in an airbridge slab so that the influence of the lightcone of this waveguide is avoided.[56] Benisty, Weisbuch (École Polytech., Benisty presently being at Orsay Univ.) et al. investigated the PBG and radiation loss mechanism in the waveguide of weak optical confinement slab with deep holes.[57] The fabrication process of this waveguide was simplified using a silicon-on-insulator wafer, and the clear light propagation with no leakage loss was demonstrated in the airbridge and some modified structures. Another type of waveguide was proposed by Yariv et al, where they noted that separated point defects can act as waveguides, so-called *coupled cavity waveguides*.[58] It is expected to be utilized as a functional device such as a dispersion compensator.

The integration of both line and point defects in photonic crystal slabs was realized by Noda et al.[59] They found that photons propagating in the line defect waveguide can be trapped by a point defect cavity placed at the vicinity of the waveguide and then emitted to vertical directions. This is physically interesting, and also provides the possibility of compact functional devices such as surface-emitting-type channel add-drop devices for optical communications. Various integrated photonic devices are now being studied worldwide, so expected to be realized shortly.

There are many photonic crystals that do not have a PBG. Kosaka et al. (NEC) suggested that such crystals could be used in an application known as a *superprism*.[60] They demonstrated a light control device by making use of this crystal's anomalous dispersion characteristic, known as a *dispersion surface*, which for photons corresponds to a Fermi surface for electrons.[61] They proposed the utilization of this property to the development of devices such as wavelength filters, dispersion compensators, and so on. Notomi (NTT) discussed the use of a uniform dispersion characteristic of a photonic crystal, which can be expressed by a *negative refractive index*.[62] He described various unique optical components based on the negative refractive index. In relation to this, discussion on the realization of *left handed light propagation,* which is realized using a combination of negative dielectric constant and negative magnetic permeability, has begun. Berger (Univ. Denis Diderot Paris) proposed the utilization of a 2D lattice to construct a nonlinear photonic crystal.[63] This idea was experimentally demonstrated in a high efficiency second harmonic generation (SHG) device.[64] These ideas may contribute towards the further expansion and design flexibility of possible device applications in the photonic crystal field as a whole.

The concept of the photonic crystal was also utilized in optical fiber technology by Russell, et al (Univ. Bath).[65] The photonic crystal fiber is manufactured with many holes arranged periodically along the light propagation direction. There are two waveguiding mechanisms; one is the PBG and the other is the total internal reflection. A fiber with an airhole core and the arrangement of holes in silica glass is known as a *PBG fiber*. It is expected to allow high power light to propagate with an extremely low loss, whilst also having low nonlinearity. A fiber with silica core and the arrangement of holes in silica is known as a *holey fiber*. Many previously unknown propagation characteristics have been discovered in this structure even though it is based on the total internal reflection.[66] It can achieve singlemode propagation in an extremely wide wavelength range, very large or small spot size, extraordinary large structural dispersion and strong birefringence. As manufacturing techniques improved, the propagation loss in holey fibers was drastically reduced to less than 1 dB/km, which is similar to that of a standard optical fiber.

Recently, many large projects are underway. In the USA, the west coast group led by Yablonovitch and the east coast group led by Joannopoulos continue to develop the field, and their projects involve many smaller groups. In Japan, three projects are underway ((i) has been finished recently): (i) A Grant-in-aid for Scientific Research on Priority Areas, from the Ministry of Education, Culture, Sports, Science and Technology. (Delegate: Inoue), (ii) A Grant-in-aid for Promotion Adjustment from the Ministry of Education, Culture, Sports, Science and Technology (Delegate: Kawakami), and (iii) Basic research supported by CREST under the auspices of the Japan Science and Technology Corporation (Delegate: Noda). In Europe, projects led by Weisbuch, Krauss, De La Rue, et al. have all commenced. The number of worldwide researchers actively engaged in photonic crystal research has dramatically increased thanks to the impact of these large projects. Recently, workshops related to photonic crystals, which were initially named WECS (Workshop on Electromagnetic Crystal Structures) and then renamed as PECS (Workshop on Photonic and Electromagnetic Crystal Structures) have been held every year: in the USA in January 1999, Japan in March 2000, in the UK in June 2001, and in the USA in October 2002. The next one will be held in Japan in March 2004. Special sessions relating to photonic crystals are now regularly being organized at major international conferences, for example CLEO, ECOC, OFC, and so on. As a result these conferences are generating great deal of interest from researchers worldwide.

References

1. K. Ohtaka, Phys. Rev. B **19**, 5057 (1979).
2. Yablonovitch, Phys. Rev. Lett. **58**, 2059 (1987).
3. S. John, Phys. Rev. Lett. **58**, 2486 (1987).
4. E. Yablonovitch, T. J. Gmitter, R. D. Meade, A. M. Rappe, K. D. Brommer, and J. D. Joannopoulos, Phys. Rev. Lett. **67**, 3380 (1991).
5. E. M. Purcell, Phys. Rev. **69**, 681 (1946).
6. K. M. Leung, and Y. F. Liu, Phys. Rev. Lett. **65**, 2646 (1990).
7. K. M. Ho, C. T. Chan, and C. M. Soukoulis, Phys. Rev. Lett. **65**, 3152 (1990).
8. E. Yablonovitch, T. J. Gmitter, and K. M. Leung, Phys. Rev. Lett. **67**, 2295 (1991).
9. J. B. Pendry, and A. MacKinnon, Phys. Rev. Lett. **69**, 2772 (1992).
10. C. M. Bowden, and J. P. Dowling, Eds., Special issue on development and applications of materials exhibiting photonic band gaps, J. Opt. Soc. Am. B **10-2**, 283 (1993).
11. K. M. Ho, C. T. Chan, C. M. Soukoulis, R. Biswas, and R. Sigalas, Solid State Commun. **89**, 413 (1994).
12. E. Ozbay, A. Abeyta, G. Tuttle, M. Tringides, R. Biswas, C. Soukoulis, C. T. Chan and K. M. Ho, Phys. Rev. B 59, 1945 (1994).
13. J. D. Joannopoulos, R. D. Meade, and J. N. Winn, Photonic Crystals, Princeton University Press (1995).
14. K. Sakoda, Phys. Rev. B **52**, 7982 (1995).
15. A. Mekis, J. C. Chen, I. Kurland, S. Fan, P. R. Villeneuve, and J. D. Joannopoulos, Phys. Rev. Lett. **77**, 3787 (1996).
16. J. D. Joannopoulos, P. R. Villeneuve, and S. Fan, Nature **386**, 143 (1997).
17. G. Tayeb, and D. Maystre, J. Opt. Soc. Am. A **11**, 2526 (1997).
18. C. C. Cheng, A. Scherer, V. A. Engels, and E. Yablonovitch, J. Vac. Sci. Technol. B **14**, 4110 (1996).
19. S. Noda, N. Yamamoto, and A. Sasaki, Jpn. J. Appl. Phys. **35**, 909 (1996).
20. P. L. Gourley, J. R. Wendt, G. A. Vawter, T. M. Brennan, and B. E. Hammons, Appl. Phys. Lett. **64**, 687 (1994).
21. T. K. Krauss, Y. P. Song, S. Thoms, C. D. W. Wilkinson, and R. M. DeLaRue, Electron. Lett. **30**, 1444 (1994).
22. T. K. Krauss, R. De La Rue, and S. Band, Nature **383**, 699 (1996).
23. T. Baba, and T. Matsuzaki, Electron. Lett. **31**, 1776 (1995).
24. K. Inoue, M. Wada, K. Sakoda, A. Yamanaka, M. Hayashi, and J. W. Haus, Jpn. J. Appl. Phys. **33**, L1463 (1994).
25. U. Grüning, V. Lehmann, and C. M. Engelhardt, Appl. Phys. Lett. **66**, 3254 (1995).
26. H. Masuda, and K. Fukuda, Science **268**, 1466 (1995).
27. M. Sigalas, C. Chan, K. M. Ho, and C. Soukoulis, Phys. Rev. B **52**, 11744 (1995).
28. J. B. Pendry, A. J. Holden, W. J. Stewart, and I. Youngs, Phys. Rev. Lett. **76**, 4773 (1996).
29. E. R. Brown, and O. B. McMahon, Appl. Phys. Lett. **67**, 2138 (1995).
30. D. F. Sievenpiper, M. E. Sickmiller, and E. Yablonovitch, Phys. Rev. Lett. **76**, 2480 (1996).

31. S. Noda, N. Yamamoto, H. Kobayashi, M. Okano, and K. Tomoda, Appl. Phys. Lett. **75**, 905 (1999).
32. S. Noda, K. Tomoda, N. Yamamoto, and A. Chutinan, Science **289**, 604 (2000).
33. A. Chutinan, and S. Noda, Appl. Phys. Lett. **75**, 3739 (1999).
34. J. G. Fleming, and S. Y. Lin, Opt. Lett. **24**, 49 (1999).
35. A. Chutinan and S. Noda, Phys. Rev. B **57**, R2006 (1998).
36. O. Toader and S. John, Science **292**, 1133 (2001).
37. S. G. Johnson, and J. D. Joannopoulos, Appl. Phys. Lett. **77**, 3490 (2000).
38. M. Notomi, T. Tamamura, T. Kawashima and S. Kawakami, Appl. Phys. Lett. **77**, 4256 (2000).
39. S. G. Romanov, A. V. Fokin, V. V. Tretijakov, V. Y. Butko, V. I. Alperovich, N. P. Johnson, and C. M. Sotormayor Torres, J. Crystal Growth **159**, 857 (1996).
40. H. Miguez, C. Lopez, F. Meseguer, A. Blanco, L. Vazquez, R. Mayoral, M. Ocana, V. Fornes, and A. Mifsud, Appl. Phys. Lett. **71**, 1148 (1997).
41. K. Yoshino, K. Tada, M. Ozaki, A. A. Zakhidov, and R. H. Baughman, Jpn. J. Appl. Phys. **36**, L714 (1997).
42. J. E. G. J. Wijnhoven, and W. L. Vos, Science **281**, 802 (1998).
43. Y. A. Vlasov, X. Z. Bo, J. C. Sturn, and D. J. Norris, Nature **414**, 289 (2001).
44. H. T. Miyazaki, H. Miyazaki, K. Ohtaka, and T. Sato, J. Appl. Phys. **87**, 7152 (2000).
45. S. Kawakami, Electron. Lett. **34**, 1260 (1997).
46. S. Shoji, and S. Kawata, Appl. Phys. Lett. **76**, 2668 (2000).
47. H. B. Son, S. Matsuo, and H. Misawa, Appl. Phys. Lett. **74**, 786 (1999).
48. M. Campbell, D. N. Sharp, M. T. Harrison, R. G. Denning, and A. J. Turberfield, Nature **404**, 53 (2000).
49. O. Painter, R. K. Lee, A. Scherer, A. Yariv, J. D. O'Brien, P. D. Dapkus, and I. Kim, Science **284**, 1819 (1999).
50. J. K. Hwang, H. Y. Ryu, D. S. Song, I. Y. Han, H. W. Song, H. K. Park, and Y. H. Lee, Appl. Phys. Lett. **76**, 2982 (2000).
51. M. Imada, S. Noda, A. Chutinan, T. Tokuda, M. Murata, and G. Sasaki, Appl. Phys. Lett. **75**, 316 (1999).
52. S. Noda, M. Yokoyama, M. Imada, A. Chutinan, and M. Mochizuki, Science **293**, 1123 (2001).
53. M. Meier, A. Mekis, A. Dodabolapur, A. Timko, R. E. Slusher, J. D. Joannopoulos, and O. Nalamasu, Appl. Phys. Lett. **74**, 7 (1999).
54. S. Fan, P. R. Villeneuve, J. D. Joannopoulos, and E. F. Schubert, Phys. Rev. Lett. **78**, 3294 (1997).
55. T. Baba, N. Fukaya, and J. Yonekura, Electron. Lett. **27**, 654 (1999).
56. A. Chutinan, and S. Noda, Phys. Rev. B **62**, 4488 (2000).
57. H. Benisty, C. Weisbuch, D. Labilloy, M. Rattier, C. J. M. Smith, T. F. Krauss, R. M. De La Rue, R. Houdré, U. Oesterle, C. Jouanin, and D. Cassagne, J. Lightwave Technol. **17**, 2063 (1999).
58. A. Yariv, Y. Xu, and R. K. Lee, A. Scherer, Opt. Lett. **24**, 711 (1999).
59. S. Noda, A. Chutinan, and M. Imada, Nature **407**, 608 (2000).
60. H. Kosaka, T. Kawashima, A. Tomita, M. Notomi, T. Tamamura, T. Sato, and S. Kawakami, Phys. Rev. B **58**, R10096 (1998).
61. P. St. J. Russell, and T. B. Birks, Photonic Band Gap Materials (Ed. by C. M. Soukoulis), Kluwer Academic, **71** (1996).
62. M. Notomi, Phys. Rev. B **62**, 10696 (2000).
63. V. Berger, Phys. Rev. Lett. **81**, 4136 (1998).
64. N. G. R. Broderiok, G. W. Ross, H. L. Offerhaus, D. J. Richardson and D. C. Hanna,

Phys. Rev. Lett. **84**, 4345 (2000).

65. R. F. Cregan, B. J. Mangan, J. C. Knight, T. A. Birks, P. St. J. Russell, P. J. Roberts, and D. C. Allan, Science **285**, 1537 (1999).

66. T. A. Birks, J. C. Knight, and P. St. J. Russell, Opt. Lett. **22**, 961 (1997).

(by Editors)

Chapter 2

OPTICAL PHENOMENA IN PHOTONIC CRYSTALS

2.1 INTRODUCTION

This chapter discusses optical phenomena in photonic crystals, which are unique and cannot be seen in ordinary materials. The essential points are outlined with examples of device applications. Therefore, readers will find some theoretical and experimental details, which have not been explained in Chapter 1. Better understanding of this chapter will be obtainable with these details described in Chapters 3 and 4.

The phenomena are divided into three categories. The first one concerns the photonic bandgap (PBG) and photon manipulation by artificial defects. It is of main interest for many researchers, since it is the fundamental principle that realizes novel nanolasers, waveguides and even photonic integrated circuits. The second one concerns the band-edge effect where the group velocity of light becomes zero. This means that a standing wave is formed over the whole photonic crystal area. Therefore, its application to large area lasers is being studied. The third one is unique characteristics of light propagation through photonic crystals, which are provided by a variety of photonic bands. Particularly, this one is further divided into two, i.e. anomalous characteristics against wavelength and incident angle of light, which is represented by the *superprism*, and the special condition near the band-edge, which allows the definition of the *negative refractive index*.

2.2 PHOTONIC BANDGAP AND DEFECT ENGINEERING

Photonic crystals are a new type of optical material with a periodic distribution of refractive index. For a 3D photonic crystal with a diamond-like structure, a PBG is formed, in which light is prohibited to propagate in any direction and are therefore completely excluded from the crystal. This is similar to the phenomenon that is found in solid crystals (especially semiconductors) with a periodic potential distribution in which an energy bandgap for electrons is formed.

For the case of semiconductors, electrons are excited from the valence band to a sufficiently high energy level in the conduction band, and are subsequently scattered mainly by phonons. In doing so, they lose energy and relax to the lowest energy level (or band-edge) in the conduction band. They then recombine with holes generated simultaneously within the valence band. In this case, the photons with an energy nearly equal to the bandgap are generated. Also, electrons (or holes), even after being accelerated by electric fields and having obtained enough kinetic energy, will be scattered by phonons and then finally relaxed to the band-edges. It means that the behavior of electrons and/or holes in semiconductors are closely related to the presence of the bandgap. Remarkable progress in optoelectronics has been developed by using these bandgap effects.

In contrast to semiconductors, phonons, which can be defined here as scattering center of photons, do not exist in photonic crystals. Therefore, when photons having an energy level equivalent to the allowed band are injected into photonic crystals, the photons will propagate through the crystals with their energy unchanged. Under such conditions, control of photons by utilizing the PBG is difficult compared to the case of semiconductors described above. When we want to use the PBG effect, it is important to construct 3D (or 2D) photonic crystals whose PBGs correspond to the wavelengths to be utilized. Also, it is necessary to introduce artificial defects to manipulate photons inside the crystals. This is known as *PBG and defect engineering*. Examples of this include light localization at point-defects or propagation through line-defects. By exploiting this type of engineering, bending of light at right angles in an ultrasmall area,[1] thresholdless lasing,[2] and even trapping of moving photons[3] should be or is possible. In the following, a representative example of PBG and defect engineering is demonstrated using photonic crystal slabs (thin films with 2D hole arrangements).

A photonic crystal slab in a triangular lattice is shown in Fig. 2.2.1(a). A line defect waveguide is formed in the crystal by missing a row of holes. The direction of the waveguide is along the Γ-J direction, which is one of two representative directions in triangular lattice. Introduction of line defect enables formation of a propagation mode in the PBG, which can be used to guide photons. Here, the PBG effect confines the photons within the line defect waveguide to in-plane directions and by the large refractive index contrast in the vertical direction. Note that a careful structural design[4] is necessary in reality for propagation without any loss in a waveguide.

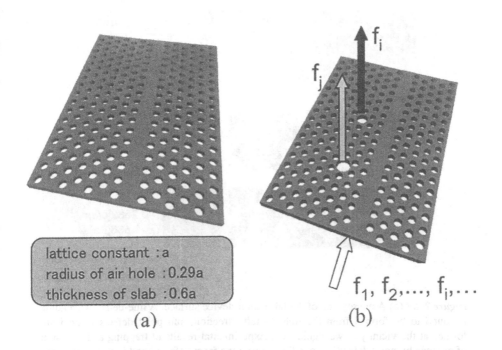

Figure 2.2.1 (a) 2D photonic crystal slab structures. The air hole radius and slab thickness are designed to be 0.29a and 0.6a, respectively, using a lattice constant of a. Waveguides with line defects are formed in the Γ-J direction in this figure. (b) Two point defects (frequencies for each defect are f_i and f_j) are introduced near waveguides of the same structure as (a).

Let us consider introducing a point-defect [3] by changing the radius of a hole at the vicinity of the waveguide, as shown in Fig. 2.1.1(b). The point-defects (named as i and j) form isolated energy levels f_i and f_j, respectively, in the PBG. When photons are introduced into the line defect waveguide, the photons with the resonant frequencies of f_i (or f_j) are trapped by the point defect i (or j). This is similar to the trapping of electrons and holes at defects in solid crystals. The trapped photons resonate inside the point defect, and during the resonance some of the photons escape from the defect in the vertical direction. In other words, point defects make it possible to trap photons propagating through the waveguide and then to emit them to vertical directions (or into free-space). This phenomenon can be applied, for example, to construct ultrasmall channel add/drop filtering devices, which are suitable for optical communications.

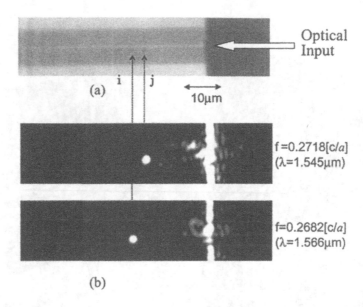

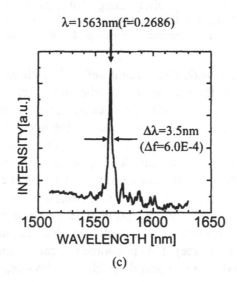

Figure 2.2.2 (a) A micrograph of the fabricated device surface. A line-defect waveguide is found to be formed from the right to left direction, and point defects i and j are formed at the vicinity of waveguide. (b) Experimental results of trapping and emission of photons by point defects. Strong light emissions from defects i and j are observed at f = 0.2718 [c/a] and f = 0.2682 [c/a], respectively.

Figure 2.2.3 (a) An example of spectra of light radiation from a point defect. The Q-factor is estimated to be about 450. (b) Relation between the defect size and the dropped wavelength.

Figure 2.2.2(a) is the top view of a fabricated sample with a line defect waveguide and point defects i and j. Photons are injected from the right-hand edge of the waveguide. When the frequency of the incident photons is equal to the resonance frequency of the defect j (f_j = 0.2718 [c/a]), a strong emission was observed from the defect, as shown in Fig. 2.2.2(b). On the other hand, when the frequency of the incident photons is changed to the resonance frequency of the other defect i (f_i = 0.2682 [c/a]), a bright emission was observed from the defect i instead of the defect j. An example of the light-emitting spectra from defects is shown in Fig. 2.2.2(c), where a distinct peak is clearly seen at a wavelength of 1.563 μm (which is equal to the resonant frequency of defect i). These results indicate a clear and concrete example of PBG and defect engineering.

When one wants to introduce many more artificial defects into the photonic crystal and construct photonic integrated circuits (or so-called *photonic chips*),[5] strong confinement of photons in the vertical direction is required. In this case, complete 3D photonic crystals[6] are essential to avoid the leakage of photons in the vertical direction. The details of 3D photonic crystals will be described in Section 3.6.

References

1. A. Mekis, J. C. Chen, I. Kurland, S. Fan, P. R. Villeneuve, and J. D. Joannopoulos, Phys. Rev. Lett **77**, 3787 (1996).
2. E. Yablonovich, J. Opt. Soc. Am. B **10**, 283 (1993).
3. S. Noda, A. Chutinan, and M. Imada, Nature **407**, 608 (2000).
4. A. Chutinan, and S. Noda, Phys. Rev. B **62**, 4488 (2000).
5. S. Noda, N. Yamamoto, M. Imada, H. Kobayashi, and M. Okano, J. Lightwave Technology **17**, 1948 (1999).
6. S. Noda, K. Tomoda, N. Yamamoto, and A. Chutinan, Science **289**, 604 (2000).

(by S. Noda)

2.3 PHOTONIC BAND-EDGE ENGINEERING

A group velocity of photons becomes zero at band-edges of photonic crystals, where standing waves are formed based on the interference (or optical coupling) among lightwaves propagating in various directions. The *band-edge engineering* described here utilizes such a standing wave. In this engineering, a complete PBG crystal is not necessary to be formed.

In a distributed feedback laser with a 1D diffraction grating, a lightwave propagating in one direction (which is defined here as the forward direction) is diffracted backward by the grating. It induces the coupling of two lightwaves propagating in forward and backward directions and a standing wave is formed, which gives 1D cavity. Band calculations reveal that the minimum loss for the cavity is obtained at two band-edges of the 1D photonic crystal. As the result, a lasing oscillation occurs at either of the band-edges. Developing this idea into photonic crystals with 2D lattices, the standing waves can be formed in 2D space through the optical coupling among lightwaves propagating in various directions in the 2D crystal. As a result, the 2D cavity modes, in which electromagnetic field distributions are completely determined by individual lattice points of the 2D crystal, can be obtained. In other words, not only a longitudinal mode (lasing wavelength) but also a lateral mode (beam pattern) can be determined by the 2D photonic crystal. Consequently, lasers which have a single mode that can oscillate coherently over a large 2D area can be produced, which makes it possible to develop novel lasers beyond conventional concepts.

An example of a 2D photonic crystal laser[1] based on the band-edge engineering is shown in Fig. 2.3.1. The device consists of two wafers A and B. Wafer A has an active layer for a gain medium, and wafer B has a triangular lattice 2D photonic crystal on top of it. Both wafers are integrated to form a laser. In this laser, a light-emitting wavelength from the active layer is designed to be equal to a period in the Γ-X direction. Note that there are six equivalent Γ-X directions. A lightwave propagating to any of six equivalent Γ-X directions is diffracted into the other five Γ-X directions, since the Bragg diffraction condition is satisfied, as shown in Fig. 2.3.2(a). The in-plane diffraction phenomena induce optical coupling among

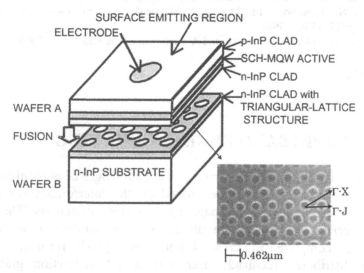

Figure 2.3.1 The structure of a 2D photonic crystal laser utilizing the band-edge effect.

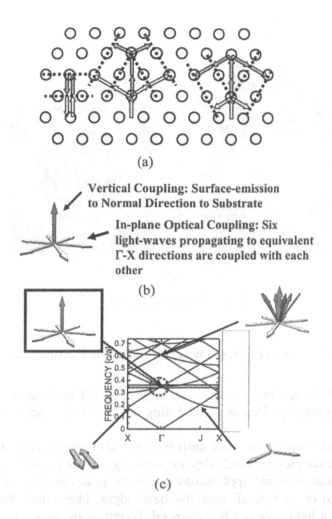

Figure 2.3.2 (a) In-plane diffracting phenomena in a triangular lattice 2D photonic crystal. (b) Optical coupling among lightwaves propagating in six equivalent Γ-X directions. The light output is emitted to the vertical direction of the plane, which means a surface-emitting function. (c) Band structures, where band-edges give the 2D cavity modes.

lightwaves propagating in the six equivalent Γ-X directions, as indicated in Fig. 2.3.2(b), which results in the formation of the 2D large area cavity mode. Since the Bragg condition is also satisfied in the vertical plane, the output power can be emitted in the direction normal to the device, as shown in Fig. 2.3.2(b). The band diagram of the structure is illustrated in Fig. 2.3.2(c). The optical coupling among the six lightwaves described above corresponds to a Γ point folded in the Γ-X direction. There are many other interesting band-edge points,[2] which give the 2D cavity modes. The top view of the actual device fabricated and the near-field pattern above lasing threshold are shown

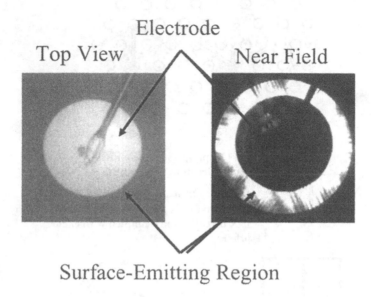

Top View
Electrode
Near Field

Surface-Emitting Region

Figure 2.3.3 Top view of the device and a near-field pattern after lasing oscillation.

in Fig. 2.3.3. It can be seen that the large area 2D oscillation has been achieved successfully. Details on the lasing properties are described in Refs. 1-3.

The discussions above have dealt with the 2D photonic crystal laser as one of the examples of band-edge engineering, although there are other attractive phenomena and applications. Since the group velocity of photons becomes zero or very small near the band-edges, interactions between a material and a lightwave can be enhanced. Electro-optic, magneto-optic, or nonlinear effects per unit length are expected to be enhanced dramatically. Some of the examples of this behavior are described in the other chapters.

References

1. M. Imada, S. Noda, A. Chutinan, T. Tokuda, H. Kobayashi, and G. Sasaki, Appl. Phys. Lett. **75**, 316 (1999).
2. M. Notomi, H. Suzuki, and T. Tamamura, Appl. Phys. Lett. **78**, 1325 (2001).
3. S. Noda, M. Yokoyama, M. Imada, A. Chutinan, M. Mochizuki, Science **293**, 1123 (2001).

(by S. Noda)

2.4 PHOTONIC BAND ENGINEERING (I)
— SUPERPRISM AND RELATED PHENOMENA —

2.4.1 The superprism phenomenon

This section describes superprism, supercollimator and superlens phenomena, which occur in a conductive photonic band that allows light transmission. The schematics of these phenomena are illustrated in Fig. 2.4.1, and typical examples are shown in Fig. 2.4.2, where the area within the dotted line was produced by the 3D photonic crystal shown in Fig. 2.4.3.[1-3]

Prisms made of glass refract sunlight into the familiar rainbow colors. The distribution of this spectrum ranges widely with wavelengths of 0.4 μm for blue to 0.7 μm for red. Despite this wide distribution of wavelength, the dispersion angle obtained is only about 10°. Consequently, about 0.1° of the dispersion angle corresponds to 1% of the wavelength difference. This compares with the refraction behavior in PCs, as shown in Fig. 2.4.2(a). Consider two light sources with wavelengths of 0.99 μm and 1.00 μm, which are chosen not only to avoid absorption by silicon but also to have a wavelength difference of about 1%. When both are injected into a photonic crystal with the same incident angle equal to 15°, the 1.00 μm wavelength light propagates almost without deviation, while the 0.99 μm light is bent through almost a right angle.[3,4] The color dispersion for every 1% of wavelength difference in this case is approximately 50°, so that the effective dispersion is about 500 times that of ordinary optical prisms. It should be noted that light with intermediate wavelengths bend at angles roughly proportional to their wavelengths, as shown in the inset of Fig. 2.4.2(a). As predicted from the figure for the 0.99 μm wavelength light, even a ±7% change of incident angle from the normal to the photonic crystal surface causes a ±70° change of propagation angle, which is a 10 fold amplification

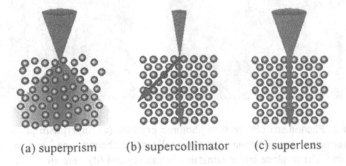

(a) superprism   (b) supercollimator   (c) superlens

Figure 2.4.1 Schematics of three phenomena obtained in photonic crystals. (a) Superprism, (b) supercollimator and (c) superlens.

of the incident angle.[2,3] This clearly shows that angular deviation of light in a photonic crystal is large for small changes in incident angle. The phenomenon was named as *superprism effect*.[2]

2.4.2 The supercollimator phenomenon

Light propagation in a *supercollimator*[3,5] has similar characteristics to those found in fiber optics, as shown in Fig. 2.4.2(c), although this is different from the waveguide effects discussed in Section 2.2. Consider a beam of light with a wavelength fixed at 0.956 μm, a diameter of 13 μm at the

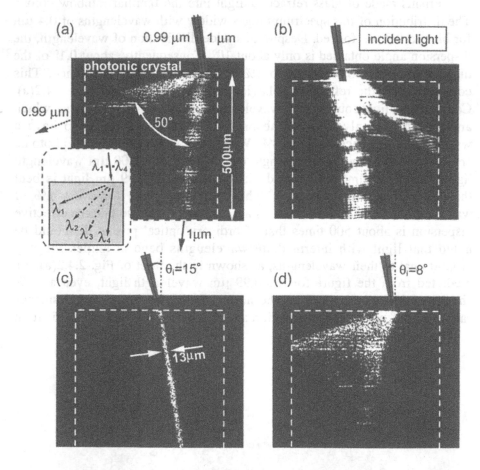

Figure 2.4.2 Phenomena observed in photonic crystals. (a) Superprism phenomenon,[2-4] (b) multi-refrigence phenomenon,[3] (c) supercollomator phenomenon,[5] and (d) superlens phenomenon.[5] The in-plane lattice constants in (a), (c) and (d) were all 0.32 μm, and 0.4 μm in (b). It was 0.32 μm in the layer normal direction for all examples. Micrographs were taken using a microscope equipped with a charge coupled detector (CCD) camera to show propagation of the laser light injected from the side of the photonic crystal.

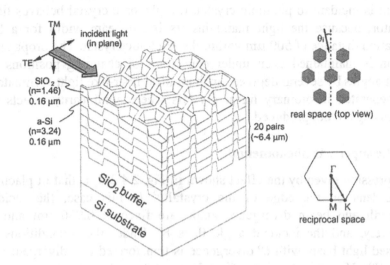

Figure 2.4.3 Photonic crystal structure used in the experiment. Within this 3D photonic crystal, the photonic atoms of Si in a medium of SiO₂ are arranged on a triangular lattice within a certain layer. By contrast, the SiO₂ photonic atoms are arranged in a triangular lattice in the Si layer in the upper layer. This 3D photonic crystal is $500 \times 500 \times 6.4 \ \mu m^3$ in size and consists of stacked layers of Si and SiO₂.

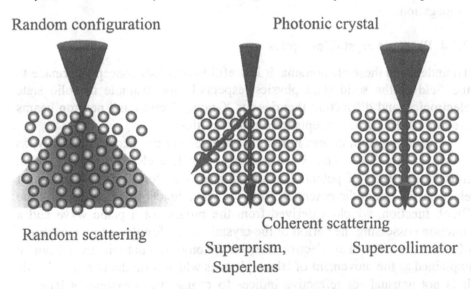

Figure 2.4.4 The comparison of ordinary scattering and scattering in photonic crystals. Scattering in photonic crystals occurs coherently. Moreover, propagation directions are divided into sensitive cases to incidental directions (these instances correspond to superprisms and superlenses) and insensitive cases (these instances correspond to supercollimators).

incident edge of the photonic crystal, and an angular divergence of 6°. When this light is incident to photonic crystal, the photonic crystal behaves like a collimator, because the light maintains its initial beam width for a long propagation distance of 500 μm within the photonic crystal. The propagation direction is maintained even under conditions of changing positions and incident angles by several degrees. Such an ideal fiber-like light collimator is never generated in ordinary linear crystals where nonlinearity effects like self-focusing are not produced.

2.4.3 The superlens phenomenon

The impression given by the effect shown in Fig. 2.4.2(d) is that of placing a concave lens on the edge of the crystal.[3,5] In this case, the incident wavelength and beam divergence angle are fixed at 0.956 μm and 6°, respectively, and the incident angle θ_i is 8°. Under these conditions the condensed light beam with 6° divergence is transformed to a divergent light beam of 70°. However, this lens effect does not depend on the position of the incident light. The numerical aperture (NA) of the light emitted out of the photonic crystal reproduces that of incident light, when the photonic crystal boundaries are aligned in parallel planes. In this case very thin beam expanders can be produced. Convex lens-like propagation is supposed to exist, but it is difficult to distinguish this behavior from concave-like propagation.

2.4.4 Photonic crystalline optics

To understand these phenomena, it is useful to note that concepts germane to the field of the solid state physics, especially for example to solid state electronics, and diffraction theories for X-rays, electron and neutron beams are more useful than concepts for ordinary optics.

Scattering from bodies in regular arrangement can occur coherently, as shown in Fig. 2.4.4. This is analogous to the free electron model, which introduces effects of potential modulation as a uniform effective mass for electrons in a periodic potential. Here the wave function is described by the Bloch function, which is derived from the product of a plane wave and a function possessing the period of the crystal lattice. Similarly to a movement of Bloch electrons in an electronic crystal, photons in photonic crystal can be explained as the movement of Bloch photons which form the photonic bands. It is not unusual for refractive indices to change from extremely large to negative values when the effective refractive indices of the photonic crystal is defined like the effective mass in an electronic crystal. This can largely exceed the value of the effective mass in a vacuum at certain times and becomes negative at other times. Actually, we can define several kinds of refractive indices in a photonic crystal, since the effective mass can be either in longitudinal or transverse directions. The three classes of refractive index are known as phase, group, and fan index.[6]

Electrons in the vicinity of a Fermi surface play an important role in the characteristics of the condensed matter of a solid. Therefore, it is generally called *Fermi surface engineering* or *band engineering*. Similarly, the photonic band structure determines the nature of photons in a photonic crystal. In the case of photonic crystals, this could be called *dispersion surface engineering* or *photonic band engineering*, whereby materials are artificially fabricated with the required optical performance inherently present and with the ability to manipulate photons as required.

Dispersion surfaces are equi-frequency contours along which the energy of the photon is conserved. The gradient direction of the dispersion surface is the direction of group velocity, which is the direction of propagation in material. This relation is written as follows,

$$v_g = \nabla_\kappa \omega(k), \tag{1}$$

where v_g is group velocity, ω is frequency and k is wavevector. Another condition is momentum conservation in the direction parallel to the incident surface. These two conservation conditions define the wavevector in a crystal and the direction normal to the dispersion surface at the wavevector point gives the direction of propagation. Typical examples of dispersion surfaces are schematically shown in Fig. 2.4.5. As the direction of the photon movement are also the gradient direction of the potential, photons

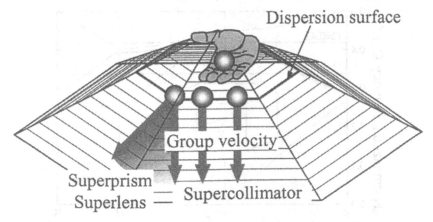

Figure 2.4.5 Conceptual figure to indicate guidelines for photonic band engineering for manipulation of light rays at will. Rays move along the potential gradient direction of an equal energy surface (dispersion surface) in photonic bands. Balls fall down in the same direction in a flat plane case, while they reflect initial conditions sensitively at angles. These behaviors correspond to the phenomenon describing supercollimators, superprisms, and superlenses.

move as if they fall down along the maximum gradient direction of this potential map. In analogy to balls falling on a flat plane, when photons are dropped, they always move in the same direction, while they alter their lateral path slightly when they are dropped on the junction between two surfaces. In other words, because of the insensitivity to the initial conditions, the former corresponds to the supercollimator phenomenon, and the latter corresponds to superprism or superlens phenomena due to its sensitivity. In this way, it is possible to manipulate light rays freely by manipulating the photonic band structure.

For the detailed analysis of superprism, supercollimator and superlens, the plane wave expansion method is normally used, which will be explained in Chapter 4. The simulated photonic band structure for the measured photonic crystals is shown in Fig. 2.4.6. Here, Ω is a normalized frequency given by $\Omega = \omega a/\pi c$ for the lateral lattice constant a. The dispersion surfaces at the frequency corresponding to the superprism have a star shape, as seen in Fig. 2.4.7. This strong modulation of the dispersion surface is the origin of the 500 times stronger dispersion behavior of superprism compared to the conventional optical gratings. Other dispersion surfaces which have rounded shapes generate different light beams in the PC. These two branches are the origin of the multi- refringence phenomenon seen in Fig. 2.4.2(a). By contrast, at the frequency of the supercollimator or superlens, the dispersion surface has both a flat range and sharp corners, which generate the supercollimated light beam and divergent light propagation, respectively, as shown in Fig. 2.4.8.

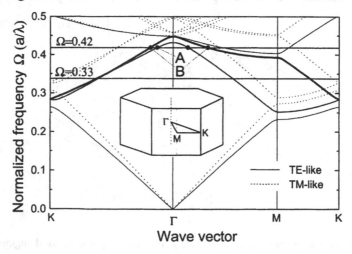

Figure 2.4.6 Photonic band structure for the measured PCs. A plane wave expansion method was used with 1258 plane waves. The solid and dashed curves denote the TE-like and TM-like modes, respectively. The inset shows wavevector paths in the first Brillouin zone.

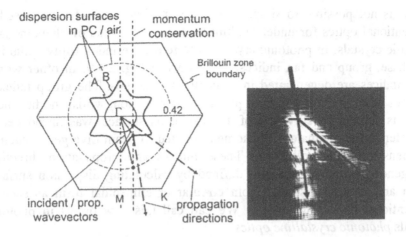

Figure 2.4.7 Dispersion surfaces folded in the first Brillouin zone at the frequency corresponding to superprism phenomena ($\Omega = \omega a/\pi c = 0.42$) for TE modes.[3] The star shaped surface is the origin of the strong dispersion (superprism). The round surface is another branch for TE modes. These branches together with TM branches are the origin of multi-refringence. The direction of propagation is normal to the surface at the intersection with the construction line (vertical dotted line). The dispersion surface is for energy conservation and the construction line is for momentum conservation. To get these dispersion surfaces, photonic bands were calculated in various directions in the K-Γ-M plain and cut at the corresponding frequencies.

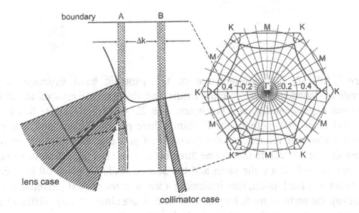

Figure 2.4.8 Dispersion surface at the frequency corresponding to supercollimator and superlens phenomena ($\Omega = \omega a/\pi c = 0.33$).[7] The point of inflection (region B) generates collimated light beam. The convex curve (region A) generates convex-like superlens phenomena. Concave-like superlens phenomena will be also appeared in region A, but hard to distinguish with convex-like propagation. The width of construction line means the spreading in *k* space because focused incident beam with 6° was used.

It is not possible to simply use the same index of refraction used in conventional optics for understanding these various kinds of phenomena in photonic crystals. In photonic crystals, the refractive index triply splits into the phase, group and fan indices,[6] as shown in Fig. 2.4.9. In other words, those indices are degenerated in conventional crystals. The group index is used for characterizing the light propagation direction, whilst the fan index which is defined by curvature of the dispersion surface, was introduced to characterize the anomalous phenomena related to beam divergence such as superlens and supercollimator. These indices and propagation direction/ divergence parameters are well defined by calculating dispersion surfaces which are strongly distorted from circular or ellipsoidal form as seen in conventional crystalline optics. We can call this new optics in photonic crystals *photonic crystalline optics.*[3]

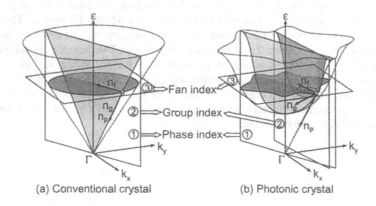

(a) Conventional crystal   (b) Photonic crystal

Figure 2.4.9   A 3D representation of the photonic band structure of (a) conventional nondispersive isotropic media and (b) photonic crystals to explain *phase index, group index,* and *fan index.* The 2D wavevector space lies in the horizontal plane that includes the Γ point. Three planes (①, ②, and ③), pass through a common reference point (stars), and produce three sections passing through the band structure that define three indices. In case (a), the sections made by planes ① and ② are the same and the section made by plane ③ is circular with a center at the Γ point, thus indicating a single conventional refractive index. In case (b), the sections made by planes ① and ② are almost always different and the local curvature of the section made by plane ③ strongly depends on the reference point that is chosen, thus creating three different indices.

2.4.5 Breakdown of classical optical theories

The following describes three unusual deviations from classical optical theories if we consider the photonic crystal as a simple classical optical element.

In classical optics, light incident on an interface between two media can be refracted and propagates through the media at an angle defined by Snell's law. That is to say that the propagation direction of the incident and refracted beam are always in the same direction. However, by comparison, consider again the 0.99 μm wavelength light incident on the photonic crystal shown in Fig. 2.4.2(a). In this case, the incident and refracted light are both on the same face of the crystal, i.e. the incident and refracted beams propagate in opposite directions. If we apply Snell's law to this effect, we obtain an effective refractive index for the photonic crystal which must be negative. For the different circumstances, it is also possible to obtain effective refractive indices which are very much smaller than 1 or extremely large, which appear to be abnormal. Clearly Snell's law is unable to predict the direction of light propagation in PCs.

Figure 2.4.2(b) shows the splitting of TE polarized light (where the electric field component is parallel to the substrate) that is incident onto a photonic crystal into two beams. The wavelength of the incident beam is 0.956 μm and angle of incidence is 8°. A TM polarized beam with the same wavelength and incident angle also splits into two beams with different propagation directions. Consequently, for incident beam, which has both TE and TM polarization components, four propagating light beams are produced. Although, at first glance, this resembles a birefringence phenomenon, birefringence never results in four rays because birefringence is defined as the splitting of one light ray into the corresponding TE and TM components. Therefore, this phenomenon is known as quadruple-refringence or multi-refringence.[3] Two branches seen in Fig. 2.4.6 are the origin of this multi-refringence. In a uniaxial crystal where birefringence occurs, one of the two resulting polarized beams must obey Snell's law (ordinary ray). However, since propagating light in a photonic crystal does not obey Snell's law, we can say that all rays in the photonic crystal are extraordinary rays.[3] Since this phenomenon is not limited to special circumstances like a vertical incidence, it is fundamentally different from the conical refraction[3] phenomenon found in ordinary optics.

Consider again the supercollimator phenomenon shown in Fig. 2.4.2(c). According to the theories of diffraction (Gaussian optics), which must be obeyed by light with Gaussian distributions, no perfectly collimated light naturally exists in a vacuum. After a collimated light beam that possesses an initial width of w_0 propagates for a distance of z, its beam width becomes w, according to the following equation:

$$w^2(z) = w_0^2(z)\left[1 + \left(\frac{\lambda_0 z}{n\pi w_0^2}\right)^2\right], \tag{2}$$

where n and λ_0 are the refractive index of the medium for propagation and the wavelength in a vacuum, respectively. The average refractive index of the photonic crystal is 2.0, being the volume average between Si and SiO$_2$. According to this equation, the beam width after a 500 µm propagation would expect to increase from 13 µm initially to 28 µm, which is greater than twice the original beam width. As far as can be judged from Fig. 2.4.2(c), the initial beam width of 13 µm does not appear to increase, which using this equation requires the use of an effective refractive index to be much greater than 10. Clearly, this phenomenon indicates that conventional diffraction theory does not apply to the light in photonic crystals. In other words, the limits imposed by conventional diffraction theory can be overcome by using photonic crystals. These apparently abnormal behaviors of photons in photonic crystals such as negative refraction, multi-refringence and diffraction free propagation are quite normal once dispersion surfaces are taken into account as explained in Subsection 2.4.4.

References

1. S. Kawakami, Electron. Lett. **33**, 1260 (1997).
2. H. Kosaka, T. Kawashima, A. Tomita, M. Notomi, T. Tamamura, T. Sato, and S. Kawakami, Phys. Rev. B. **58**, R10096 (1998).
3. H. Kosaka, T. Kawashima, A. Tomita, M. Notomi, T. Tamamura, T. Sato, and S. Kawakami, J. Lightwave Technol. **17**, 2032 (1999).
4. H. Kosaka, T. Kawashima, A. Tomita, M. Notomi, T. Tamamura, T. Sato, and S. Kawakami, Appl. Phys. Lett. **74**, 1370 (1999).
5. H. Kosaka, T. Kawashima, A. Tomita, M. Notomi, T. Tamamura, T. Sato, and S. Kawakami, Appl. Phys. Lett. **74**, 1212 (1999).
6. H. Kosaka, A. Tomita, T. Kawashima, T. Sato, and S. Kawakami, Phys. Rev. B **62** 1477 (2000).
7. H. Kosaka, Physics (Buturi) **55**, 172 (2000, in Japanese).

(by H. Kosaka)

2.5 PHOTONIC BAND ENGINEERING (II)
   — 'REFRACTIVE' PHENOMENA —

2.5.1 Diffraction or refraction

As described in Section 2.4, the interesting aspects of photonic crystals are not limited to responses within PBG wavelengths. Photonic crystals work as good conductors for light wavelengths in photonic bands (that is to say, outside the PBG range), and these band structures determine their optical propagation characteristics. In other words, the optical propagation characteristics of photonic crystals can be designed using the crystal structure. The possibility for producing novel dielectric materials, possessing new optical characteristics, using photonic crystals, has therefore been proposed.[3,4]

Unfortunately this idea is not as easy to achieve as it seems. Indeed, optical propagation characteristics in photonic crystals are quite different from ordinary dielectrics like glass. For example, the widely held Snell's law, which is useful for ordinary optical materials, becomes invalid as described in previous section. However in many cases, propagation characteristics can be understood as diffraction phenomena caused by a grating. This is particularly the case for photonic crystals with a weak refractive-index modulation, that behave in an almost equivalent manner to the classical diffraction grating theory. The aim of this section is to describe the behavior of photonic crystals with a large refractive-index modulation and show whether or not the propagation characteristics will be essentially different from diffraction gratings.[5,6] In a sense, for photonic crystals, the classical diffraction grating theory can be replaced by electronic band theories.

To begin with we explain optical propagation characteristics in ordinary dielectrics and diffraction gratings from the point of view of photonic crystals. Then the propagation characteristics of photonic crystals with weak refractive index modulation will be reviewed and compared with diffraction gratings. It will be shown that the propagation characteristics in photonic crystals with strong refractive index modulation are quite different from classical diffraction gratings and are in fact rather similar to refractive phenomena in dielectrics. The propagation characteristics follow Snell's law which determines propagation paths in dielectrics and they can concurrently possess an extraordinary refractive index value (a negative value or value less than 1). In other words, such photonic crystals behave as dielectrics possessing 'artificial' refractive indices concerning beam propagation. The mediums possessing such negative refractive indices have interesting imaging effects and have the potential to allow us to implement totally new optical devices.

2.5.2 Optical propagation in dielectrics, diffraction gratings, and photonic crystals

As shown in Fig. 2.5.1(a), parallel beams are injected from one medium, air in this example, of refractive index n_1, to a dielectric medium of refractive index n_2. The relation between the incident angle and the propagating angle are described by Snell's law for ordinary dielectrics as

$$n_1 \sin \theta_1 = n_2 \sin \theta_2. \tag{1}$$

Snell's law allows for a wavenumber conservation law of lightwaves at the

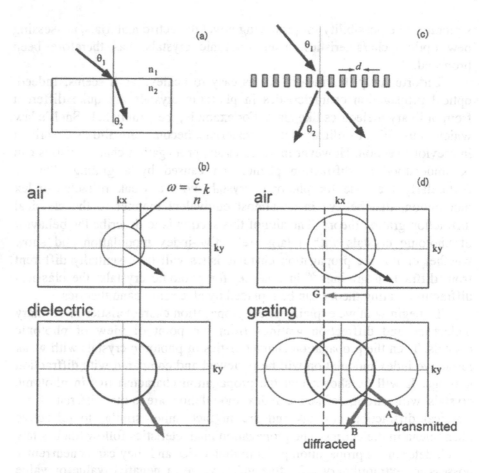

Figure 2.5.1 (a) Beam incidence from air to an ordinary dielectric. Refraction occurs in this case according to the refractive index value of the dielectric. (b) The wavenumber conservation law at the interface can be illustrated in diagrams using an equi-frequency surface in the wavenumber space. (This is Snell's law diagrammatically illustrated.) (c) Incidence from air to the diffraction grating. (d) The equi-frequency surface becomes a repetition of circles in this case.

interface, which can be illustrated as in Fig. 2.5.1(b). It is the so-called elliptic expression of refractive index that corresponds to describing an equi-frequency surface of photonic bands in the wavenumber space (bands in dielectrics are represented as $\omega = (c/n)k$). Illustrating the wavenumber conservation law using diagrams of wavenumber space is now widely accepted.

Now, consider the case of beam injection into a 1D diffraction grating as shown in Fig. 2.5.1(c). The periodicity of the diffraction grating, as seen in Fig. 2.5.1(d), shows that an equi-frequency surface or dispersion surface

becomes a repetition of a circle. As a result, the wavenumber conservation
lines have intersecting points with more than one circle, which correspond to
the transmitted beam (0th-order diffracted beam) and the diffracted beams.
The beam propagation direction in this case obeys the following well-known
diffraction law, which is illustrated in Fig. 2.5.1(d),

$$m\lambda = d \sin(\theta_1 + \theta_2),\qquad(2)$$

where m is the integer number, λ wavelength, and d is the lattice period. The
propagation problem in photonic crystals is considered in a similar way.
However, the equi-frequency surface is determined by the band structure.[7]
For triangular lattice 2D photonic crystals composed of cylinders, as shown
in Fig. 2.5.2(a), the first Brillouin zone of the reciprocal lattice space with
the same structure is shown in Fig. 2.5.2(b). The equi-frequency surface
where the refractive index modulation effect is weak is shown in Fig.
2.5.2(c). In this figure it can be seen that there are two independent points
satisfying the wavenumber conservation law, and propagation directions are
determined by normal lines on the equi-frequency surface at those points.
Note that the direction is always outward-going concerning each equi-
frequency circle, which is the same as the situation in homogeneous media
and diffraction gratings. Although the shape of this equi-frequency surface at
first glance looks complex, it is obvious when comparing Fig. 2.5.1(d) to Fig.

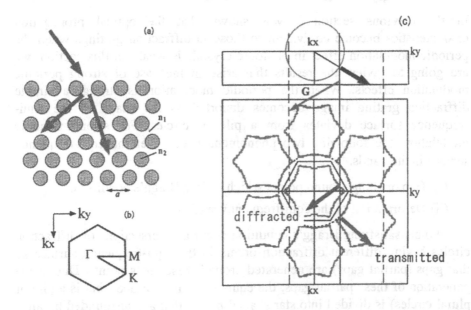

Figure 2.5.2 A beam incident from air to a photonic crystal with weak modulation. (a)
Crystal structure is a 2D cylindrical triangular lattice. (b) The first Brillouin zone of this
photonic crystal. (c) The equi-frequency surface for the weak modulation case.

2.5.2(c) that photonic crystals resemble diffraction gratings, in that the circles are centered at the origin and repeatedly reflect the periodicity of the crystal. The two beams induced in the crystal are shown in Fig. 2.5.2(c), and correspond to the transmission and diffraction waves. In photonic crystals and 1D and 2D diffraction gratings,[3,8,9] a beam-steering phenomenon occurs whereby the change of incidental angle causes a large change in propagation direction angle. This can be understood by observing that the change of angle occurs in the vicinity of the intersecting points of the circles when an excited wave moves to another circle of different order by changing the incidental angle.

Therefore, light propagation phenomenon in weakly-modulated photonic crystals is very different from refraction phenomenon in Fig. 2.5.1(a) and (b) and is basically *diffraction* phenomenon[10]. The simple picture (such as Snell's law) for conventional *refraction* phenomena is difficult to be applied, and the definition of refractive index becomes vague in weakly-modulated photonic crystals. However, this situation will fundamentally change if we consider strongly-modulated photonic crystals, which we will see in the following parts.

2.5.3 Optical propagation in photonic crystals with a strongly modulated refractive index

In the previous section it was shown that the optical propagation characteristics become equivalent to those in diffraction gratings when the periodic modulation effect in photonic crystals is weak. In this section, we are going to review the results that arise in the case of strong periodic modulation effects. When the periodic modulation is strong, a simple diffraction grating image becomes distorted as the shape of the equi-frequency surface deviates from a pile of circles. For strong periodic modulation, the following two phenomena occur over the equi-frequency surface in the bands.

(i) Gap opening around points which satisfy Bragg's condition.

(ii) Deformation of the equi-frequency surface.

Points satisfying Bragg's condition occur at intersections of diffraction circles having different diffraction orders in the equi-frequency surface so that gaps (partial gaps) are generated around these cross points. Due to the generation of these partial gaps, the equi-frequency surface (that is a pile of plural circles) is divided into star shaped pieces that are surrounded by arcs. Partial gaps are generated, even in photonic crystals of weak modulation and diffraction gratings, though their sizes are negligible. Periodic modulation effects appear and begin to affect the whole equi-frequency surface by expanding the partial gaps. Finally, these partial gaps expand to whole wavenumber space to form full gaps, which are called PBGs.

The equi-frequency surface inevitably deviates from a simple pile of circles and loses its shape during the expansion process of the partial gaps. The degree of deviation becomes greater as this shape becomes closer to the points which satisfy Bragg's condition (that is to say, crosses points of the circles). As sharp vertices of star-shaped pieces are nearest to these cross points, the sharp vertices of the equi-frequency surface are deformed into shapes with rounded vertices. As a result, the equi-frequency surfaces become circle-like (or sphere-like) shapes around certain symmetrical points (the center surrounded by points which satisfy Bragg's condition is always a symmetrical point in a crystal) according to the degree of periodic modulation effects.

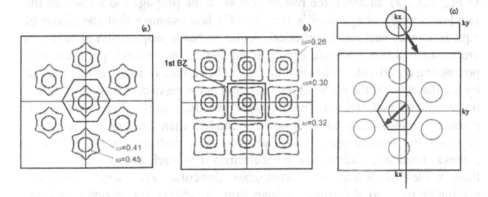

Figure 2.5.3 The equi-frequency surface in wavenumber space regarding 2D photonic crystals of a cylindrical triangular lattice (a) and 2D photonic crystals of a cylindrical square lattice (b). The ω is a normalized frequency. The shape of the equi-frequency surface is star-shaped or rectangular when the frequency is far from the gaps (the modulation of refractive index is weak), while the shape becomes round as the frequency approaches the gaps (at the nearest point $\omega = 0.41$ for (a), and $\omega = 0.32$ for (b)). (c) A conceptual figure of the equi-frequency surface for beam incident onto a photonic crystal with strong index modulation.

The equi-frequency surfaces for 2D photonic crystals composed from a cylindrical triangular lattice and square lattice photonic crystals are shown in Figs. 2.5.3(a) and (b). Frequencies are normalized by the crystal period, *a*. When the frequency approaches a corresponding gap energy, the vertices of the equi-frequency surfaces have a tendency to be rounded for both triangular and square lattices. However, when the frequency is far from the gap and the periodic modulation effect is weak, the shapes of the equi-frequency surfaces reflect the original symmetry and keep their star-like or rectangular shapes. The equi-frequency surface finally becomes a complete circle shape in the vicinity of the gaps, an effect well-known in solid state physics. The Fermi surface in the band theory of solid state physics (which determines the characteristics of a metal) corresponds to the equivalent surface for frequency. The shape of the Fermi surface is star-like under weak periodic modulation and then changes to a circle under strong periodic modulation near the gaps.[11] Physically this is the same phenomenon as the above-mentioned phenomenon for photonic crystals.

The fact that the equi-frequency surface becomes rounded leads to results. Figure 2.5.3(c) shows the situation where light of frequencies near the gap are injected into 2D photonic crystals with a triangular lattice structure. A single wave is excited in the crystal in this case, which is determined only from the central circle. Therefore, when the equi-frequency surface becomes round, this beam injection becomes equal to the dielectric in Fig. 2.5.1(a). In short, the relation between the propagation angle and the incidental angle obeys Snell's law. Snell's law assumes that the shape of equi-frequency surfaces is spherical. As far as beam propagation phenomena are concerned this photonic crystal is equivalent to an isotropic dielectric, possessing a particular refractive index.[12] The refractive index is determined by the diameter of the circle in this case and is not restricted by the refractive index of the material but by the band structure itself. This means that refractive indices can have values far smaller than one, or can even be negative. Put another way, if the equi-frequency surfaces are not circles or spheres, refractive indices that are determined in such a way, may change their values according to wavenumber direction. As Snell's law was originally used to determine propagation directions for various incident directions at a constant refractive index, the acquired refractive indices make no sense in these cases. On the other hand, Snell's law becomes meaningful when the frequency approaches the gap and the defined refractive indices begin to fit.

2.5.4 Comparison with band theory in solid state crystals

The band structure of a 2D photonic crystal of cylindrical triangular lattice is shown in Fig. 2.5.4(a).[13] In this band, a small gap opens at Γ_3 ($\omega = 0.635$) point. The frequency dependencies of the effective refractive indices (determined at points near the small gap by the method introduced in the previous subsection) are shown in Fig. 2.5.4(b). In this figure, the area indicated by the solid lines connecting the data, show the frequency range where the equi-frequency surface shape is a circle, and accordingly, where refractive indices are meaningful. Provided that the wavelength at the gap center is 1.5 μm, the effective wavelength range for effective refractive indices is over 100 nm, which indicates that the concept is effective over a

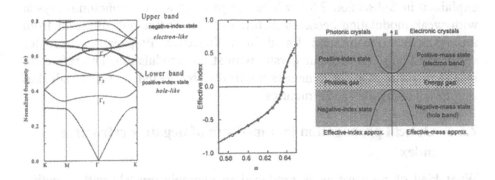

Figure 2.5.4 Band structures of a 2D photonic crystal of column type Si triangular lattice for waves polarized in the horizontal direction. (b) Frequency dependencies of effective refractive indices near the Γ_3 point. (c) Comparison of the relationship between an effective index and a band in a photonic crystal, and the relationship between an effective mass and a band in a semiconductor.

considerably broad wavelength range. As is clear from Figure 2.5.4(b), the gap position is located between the positive and negative state of the refractive index. Positive indices appear in the upper region of the gap and negative refractive indices are in the lower region of the gap. A negative refractive index corresponds to a negative band gradient, which shows the mechanism's similarity to semiconductor crystals where an electron band having positive effective mass is generated at the top of the bandgap and an electron band having negative effective mass is generated at the bottom of the bandgap. (In both cases the effective masses are determined by the band gradients.) In general, in bands of electrons in a solid state crystal, electron bands with positive effective masses and hole bands with negative effective masses are formed at the upper and lower regions of the bandgap, respectively. It is interesting that photonic crystals and ordinary electronic crystals correspond to each other in their behavior.

States, which can be expressed by using effective refractive-index, are generally feasible in a region of strong periodic modulation in a photonic crystal, in a wavelength range near a gap. In other words, photonic crystals with strong PBGs work as isotropic media possessing virtual refractive indices in a wavelength region in the vicinity of the gaps. This phenomenon is analogous to the case for band electrons. Bloch (band) electrons near gaps in a semiconductor crystal move freely in spite of complex scattering from the crystal lattice and they are regarded as free electrons with certain effective masses. In the same way, photons near the PBGs can be understood to propagate in the same manner as free photons possessing constant refractive indices in a crystal.

Optical propagation in a weakly-modulated photonic crystal is explained in Subsection 2.5.1 where the propagation in a photonic crystal with weak modulation becomes diffraction grating-like and different from the refractive phenomenon found in a dielectric. Propagation becomes *refractive* again in a photonic crystal with strong modulation. The refractive index controlling the phenomena is not limited by the index of materials, but is determined by the band structures.

2.5.5 Optical propagation in a medium of negative refractive index

What kind of propagation is produced in photonic crystals with negative refractive indices? As no material exists which has a negative refractive index, an extraordinary propagation effect occurs that is never achieved under normal conditions. Examples of optical propagation with unusual characteristics that can be achieved in media of negative refractive indices are shown in Fig. 2.5.5. All of these examples are propagation phenomena which can be induced simply by hypothesizing a negative index in the normal Snell's law scenario.

Optical propagation in such a state is investigated in this subsection by means of numerical simulation using the finite differential time domain method (FDTD method).[14] In the FDTD method, we directly discretize Maxwell's equations in space and time, to obtain time evolution of the lightwave field in photonic crystals as explained in Chapter 4. This method corresponds to a numerical analysis that is independent of band calculations, and allows the possibility of confirming phenomena predicted from band calculations.

In simple terms, negative refractive indices mean that negative values are derived from Snell's law. In other words, it corresponds to the state where the wavevector direction of a light beam is opposite to the group velocity vector direction. Such a refraction, which cannot occur in an

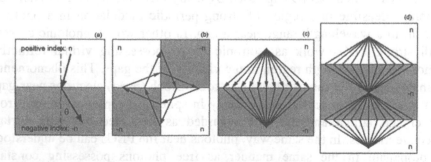

Figure 2.5.5 Examples of propagation in mediums with negative refractive indices.

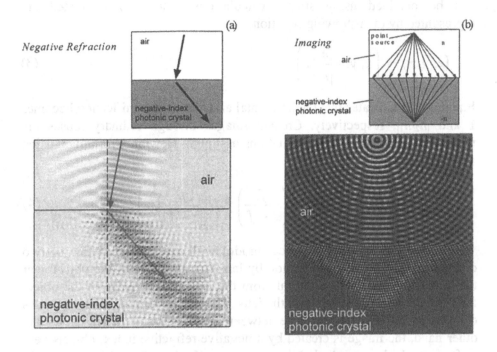

Figure 2.5.6 Simulated result for the propagation of light injected in a 2D triangular lattice photonic crystal of negative refractive index. The refraction of parallel beams incidence to a negative angle. (b) The propagation of beams incident from a point optical source in a photonic crystal. Image creating actions are carried out by the focusing.

ordinary case, is actually realized in a photonic crystal. Results of numerical calculation by FDTD for beams incident from air into 2D cylindrical triangular lattice photonic crystals of a negative refractive index state are shown in Fig. 2.5.6. It can be seen that excited waves propagate in the negative direction of the refractive index in the crystal. This refraction angle agrees with the propagating angle, calculated from Snell's law, using the negative effective refractive index acquired from band calculations.

Among the several interesting optical propagation phenomena induced from such an extraordinary *negative refraction*, the image creation phenomenon is extremely interesting, and is explained as follows. Consider the hypothetical situation where two mediums exist, one with a negative refractive index, n_2 (< 0), and the other with a positive refractive index, n_1. Light from a point source in the n_1 medium propagating into n_2 can be understood by use of the para-axial process used in classical geometric optics. This is the same as treating images created by lenses. In this case it is found that the image creation in a medium with a negative refractive index can be obtained using simple calculations. The images created are represented by the following equation,

$$(x, y, z) \rightarrow \left(x, y, -\left|\frac{n_2}{n_1}\right| z \right).$$
(3)

Equation (3) indicates that the horizontal and vertical magnification becomes 1 and $|n_2/n_1|$, respectively. Created images through ordinary lenses are represented by the following equations assuming Newton's formulae, using a focal distance f,

$$(x, y, z) \rightarrow \left(-\left(\frac{f}{z-f}\right)x, -\left(\frac{f}{z-f}\right)y, -\left(\frac{f}{z-f}\right)^2 z \right).$$
(4)

Figure 2.5.7 shows a typical model with the characteristics of two created images. For image creation by lens (b), an image of the object near the lens is created at the point far from the lens and an image of the object far from the lens is created near the lens. The magnification power of lenses changes according to the distance between the objects and the lenses. On the other hand, the image is created by a negative refractive index, objects near or far from the lens create their images near or far from the lens, respectively, showing that magnification power has no relation to distance. That is to say, half of the whole infinite space which is on the side of a positive refractive index creates its reflected image in the opposite space in the direction of the

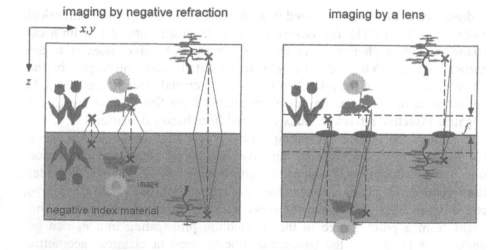

Figure 2.5.7 (a) Image creation in a negative refractive index medium. (b) Image creation with an ordinary lens.

z-axis. Image creation by a negative refractive index can be considered as 3D space imaging, while image creation by lenses is regarded as creation in the 2D plane. In this sense, the image created by a negative refractive index is closer to an image created by a mirror. However, there is a large difference between them in that the real images are created by a negative refractive index, compared to the virtual images created by mirrors.

Light propagation from a point source in a medium with positive refractive index to a photonic crystal with a negative refractive index is simulated by the FDTD method. The supposed photonic crystal structure is the same as that in Fig. 2.5.3, and light with ω = 0.62 is launched. As shown in Fig. 2.5.6(b), the light is focused on one point of the photonic crystal in a negative refractive index state. The position of this image agrees with the position acquired from geometric optics considering a negative refractive index, which is the real image formation mechanism in a photonic crystal.

2.5.6 Negative refractive indices in periodic structures of metals and relation with photonic crystal

The phenomena of negative refractive indices near the PBG of dielectric photonic crystals has been explained in previous sections. At about the same time as the publication of this phenomena,[5,6] the possibility of negative effective refractive indices in metal periodic structures was pointed out and then actually observed by experiment. Consequently, negative refractive indices are attracting worldwide attention.[15] Although this negative refraction phenomenon in metal periodic structures resembles that of photonic crystals in terms of controlling the effective refractive indices by photonic crystal structure, the principles are different from each other in the following ways.

(i) It occurs in a wavelength range that is far longer when compared to the period. (This is the so-called effective medium range.)

(ii) It utilizes a dispersion relation which is characteristic to metals.

(iii) As a consequence of points (1) and (2), the effect is only produced in the microwave range.

Since Ref. 15 describes this area in detail, only a brief description of this behavior will be given here.

The fundamental principle dates back to a paper by Veselago in 1968.[16] Producing media with negative refractive indices and the possibility for propagation of an electromagnetic wave through such media were theoretically predicted assuming simultaneous negative values of a dielectric constant and a magnetic permeability. This prediction has not actually been achieved experimentally because of the difficulty in changing the magnetic permeability in ordinary materials. Recently, however, the effective magnetic permeabilities (and dielectric constants) have been shown to be controllable by the metal periodic structures of certain geometric structures.[17] In addition, using such structures, realization of Veselago's prediction has been shown to be probable for a metal system, by periodic arrangement of structures having either negative magnetic permeability or a negative dielectric constant.[18] Soon after this discovery, a structure for operation in the GHz region was fabricated in a trial to experimentally confirm the negative refractive index.[19]

As these various attractive negative refraction phenomena (such as the image creation phenomena described above) do not depend on the physical mechanism of negative refractive indices, the same phenomena can be achieved in these metallic structures. The frequency range for negative refractive indices of metal periodic structures is restricted to within the range where magnetic permeability becomes negative, while the frequency range for effective negative refractive indices is limited to the vicinity of PBGs in the case of dielectric photonic crystals. The most attractive subject concerns the highest limit of the frequency range of this phenomenon. Although the permeability must approach a value of 1 even in periodic structures, because of the magnetic permeability, it is generally known to approach unity in the higher frequency range.[20] However, there seems to be no clear answer concerning the limits of high frequency range.

Concepts to produce artificial dielectrics using photonic crystals are strongly related to the idea of metamaterials, which is based in the field of periodic structures of metal systems. The importance of the field is emphasized by the fact that specialized international conferences have been organized on metamaterials, covering both the fields of photonic crystals and metal periodic structures.[21] Metamaterial research represents the intention to produce materials possessing new optical or electromagnetic properties, by the use of artificially fabricated structures that are larger than molecular or atomic structures. This field is very closely related to the photonic crystal field, although this closeness is not always representative because Bragg reflection is not necessarily used. Interactions between the world of dielectric photonic crystals (concerning optical fields) and the world of metal periodic structures (concerning electromagnetic fields) are destined to become stronger. This is because various photonic crystal applications, designed to function outside the PBGs, are being considered and there appears to be a movement to recognize photonic crystals as new bulky optical materials.

References

1. J. D. Joannopoulos, R. D. Meade, and J. N. Winn, Photonic Crystals, Molding the Flow of Light, Princeton University Press (1995).
2. C. M. Soukoulis Ed., Photonic Band Gap Materials, Plenum Press (1995).
3. P. St. J. Russell, T. A. Birk, and F. D. Lloyd-Lucas, Confined Electrons and Phonons (Ed. by E. Burstein, and C. Weisbuch), Plenum Press (1995).
4. K. Ohtaka, T. Ueta, and Y. Tanabe, J. Phys. Soc. Jpn. **65**, 3068 (1996).
5. M. Notomi, Phys. Rev. B **62**, 10696 (2000).
6. M. Notomi, Proc. SPIE **4283**, 69 (2001).
7. M. Notomi, T. Tamamura, H. Ohtera, O. Hanaizumi, and S. Kawakami, Phys. Rev. B **61**, 7165 (2000).
8. H. Kosaka, T. Kawashima, A. Tomita, M. Notomi, T. Tamamura, T. Sato, and S. Kawakami, Phys. Rev. B **58**, R 10096 (1998).
9. R. Zengerle, J. Mod. Opt. **34**, 1589 (1987).
10. It is difficult to distinguish strictly the photonic crystal and a diffraction grating. Here, the characteristic of diffraction grating is considered to occur from the band folding by the periodicity. In photonic crystals, not only the band folding but also the opening of the PBG occur, and peculiar effect appear not only at the Bragg frequency but also at other frequencies. The generation of diffracted wave, which is expressed by the formula, is due to the band folding.
11. J. M. Ziman, Principles of the theory of solids, Cambridge University Press (1972).
12. The refractive index defined here is an index that determines the refraction phenomenon described by Snell's law. So, it corresponds to the phase index of a dielectric medium, and does not the group index given by the group velocity of light.
13. The band structure and dispersion surface were calculated by the plane wave expansion method.
14. A. Taflove, Computational Electrodynamics: The Finite Difference Time Domain Method, Artech House (1995).
15. D. R. Smith, W. J. Padilla, D. C. Vier, R. Shelby, S. C. Nemat-Nasser, N. Knoll, and S. Schultz, Photonic Crystals and Light Localization in the 21st Century (Ed. by C. M. Soukoulis, Kluwer Academic (2001).
16. V. G. Veselago, Soviet Physics USPKHI, **10**, 509 (1968).
17. J. B. Pendry et al., IEEE Trans. MTT, **47**, 2075 (1999).
18. D. R. Smith, W. Padilla, D. C. Vier, S. C. Nemat-Nasser, and S. Schultz, Phys. Rev. Lett. **84**, 4184 (2000).
19. R. A. Shelby, D. R. Smith, and S. Shultz, Science **292**, 77 (2001).
20. L. D. Landau, and E. M. Lifshitz, Electrodynamics of Continuous Media, Pergamon Press (1985).
21. Progress in Electromagnetics Research Symposium, Cambridge, Massachusetts, June (2002).

(by M. Notomi)

References

Chapter 3

THE PRESENT STATUS OF PHOTONIC CRYSTAL DEVELOPMENT

3.1 INTRODUCTION

Photonic crystal research is advancing at an astonishing pace. In this chapter photonic crystals are classified, from a crystal dimension viewpoint, into 2D and 3D crystals. They are further divided, from a material point of view, into semiconductors, organic dielectrics and fibers. Although 1D crystals have occasionally been referred to in recent photonic crystal research, they are omitted from this chapter.

First, photonic crystal fabrication research is classified and listed in a table in Section 3.2. According to the classification above, the present status of crystal fabrication techniques, an overview of the present way of thinking and a look into the future of photonic crystal research are introduced in Sections 3.3 – 3.11. Note that some overlapped descriptions and different terminologies will appear in these sections since the individual authors have written separate parts independently; however, we hope that this will give an interesting variety to our explanations of the concepts involved in photonic crystal research.

3.2 SUMMARY OF STRUCTURES AND RESEARCHES

Research groups, representative papers, fabrication techniques, materials, research achievements, and future prospects are summarized in Table 3.2.1, where only research groups who *fabricated* such structures are listed. Not only those for lightwaves but also for microwaves and millimeter waves are included, whilst those for acoustic waves and elastic waves are omitted since they are not electromagnetic waves. The table shows that the research field is growing for the purpose of various wave controls.

(by Editors)

1) Organizations are listed with A,B,C--- order. One publication is selected for each organization.Following abbreviations are used; AO: Appl. Opt., APL: Appl. Phys. Lett., EL: Electron. Lett., JAP: J. Appl. Phys., JJAP: Jpn. J. Appl. Phys., JLT: J.Lightwave Technol, JQE: IEEE J.Quantum Electron., JSTQE: IEEE J.Selected Topics in Quantum Electron., JVST: J. Vacuum Sci. Technol, MTT: IEEE Trans. Microwave Theory & Technol., NL: Nano Letters, NT: Nature, OE: Opt. Express, OL: Opt. Lett., OSA: J.Opt.Soc.Am B, PRB: Phys. Rev. B, PRL: Phys. Rev. Lett., PTL: IEEE Photon.Technol. Lett.. SC: Science, TAP: IEEE Trans. Antennas & Prop. Numbers are volume and the first page (or issue number).

2) Following abbreviations are used; AN: anodization, BN: bonding, BS: bias sputter, DE:dry etching, DF: diffusion, DP: deposition, EP: epitaxy, EV: evaporation, FB: Fiber drawing, IO: inverse opal, MF: microfabrication, MN: manipulation, MPA: multi-photon absorption, OX: selective oxidation, PS: polish, SO: self-organization, WE: wet etching, XL: X ray lithography, IL: interference lithography, FL: fs exposure, PL: poling.

3) GaAs, InP, etc. includes related compounds.

4) (G), (E) and (B) denote those related with PBG and defect, band-edge, and photonic band effects, respectively. Here, PBG includes stop band for only one direction.

Dimension	Structure	Organization (Publication[1]) (A,B,C --- order)	Fabri-cation[2]	Material[3]	Activity[4]	Prospects, etc.[4]
2-D	Deep holes	Caltech (EL32,2243) Ecole Polyt. (JQE35,1045) Glasgow U. (NT383,699) Hokkaido U. (JJAP33,L1463) Max-Planck-Inst. (APL78,1180) Nagoya U. (JJAP36,7763) NTT (JJAP39,6259) Photon. Nanostr. Lab. (OL26,1259) Sandia Nat'l Lab. (APL64,687) Siemens (APL66,3254) THOMSON-CSF (EL33,no5) Tokyo Metro. U. (JJAP37,L1340) U. Konstanz (PRB64,233102) U. Michigan (JQE37,1153) U. Paris (JLT17,1989) U. Pavia (PRB65,112111) U. Sheffield (JLT17,2050) U. Southampton (APL67,1877) U. Tokushima (OSA18,1084) U. Toronto (APL75,3063) U. Wisconsin (OL26,1353) U. Würzburg (APL79,4091)	DE AN	GaAs InP SiO₂ Si Al₂O₃ LiNbO₃	(G) Evaluation of PBG (G) Laser mirror (G) Observation of defect mode	(G,E) Lasing in point defect and band-edge (G) Light propag. in waveguide with bend, etc. (B,E) Observation of strong dispersion characteristics (B) High efficiency LED (B,E) Large nonlinear device
	Slab	Caltech (SC284,1819) CNRS (APL80,547) FESTA (APL79,4286) Hokkaido U. (PRL86,2289) KAIST (APL76,2982) Kyoto U. (NT407,608) Lyon Ecole Center (EL37,no12) NEC (APL76,952) NTT (EL37,no5) Sandia Nat'l Lab. (NT407,984) Technical U. Denmark (EL38,no6) Tech. U. Hamburg (APL78,2434) U. Bath (JAP85,6337) U. Bourgogne (OL27,173) UCLA (APL75,1036) UCSB (APL78,2279) U. Montpellier (JLT17.2058) U. Paris (EL37,no5) USC (PTL14,435) U. Science and Technologies of Lille (EL35,no6) U. Sheffield (OSAB19,716) U. Southampton (APL76,991) U. Tsukuba (JAP89,855) U. Würzburg (EL38,no4) Yokohama Nat'l U. (EL27,654)	DE	InP GaAs Si Si₃N₄ Polymer	(G) Evaluation of PBG (G) Lasing in point defect (E) Operation in 2-D DFB laser (G) Light propag. in waveguide with bend, etc. (G,E) Observation of ultra-small group velocity (G) Demonstration of defect filter (B) Observation of high light extrac. efficiency	(G,E) Low threshold lasing (G) Large scale high density functional photonic circuit (G) All optical switch (B) High efficiency LED (B) Superprism filter, light deflector, polarizer (B,E) Delay line, disper compen -sator (B,E) Large nonlinear device

Dimension	Structure	Organization (Publication[1]) (A,B,C --- order)	Fabrication[2]	Material[3]	Activity[4]	Prospects, etc.[4]
2-D	Pillars	FOM Inst. Atomic and Molecular Phys. (JVST17,2734) Glasgow U. (EL30,1444) Inst. Phys. and Chem. Res (JJAP36,L286) JRCAT (APL75) NTT (JJAP39,6259) U.S. Army Res. Lab. (JAP82.6354) U. Tokushima (OSA18,1084) Yokohama Nat'l U. (JLT17,2113)	DE EP	GaAs InP Si Glass	(G) Evaluation of PBG (G) Laser mirror (G) Light propag. in waveguide (B) Observation of high light extrac. efficiency	(G) Low loss waveguide (B) High efficiency LED
	Surface Grating	Caltech (JQE36,1131) Harvard U. (AO38,5799) MIT (APL78,563) Lucent Technologies (APL74,7) NTT (APL78,1325) Pennsylvania State U. (APL79,3392) U. München (APL77,2310) U. Southampton (PRL84,4345)	DE PL	GaAs InP LN Dye Polymer Ag	(E) Operation in 2-D DFB laser (B) High efficiency LED (B,E) High efficiency SHG (B) Wavelength conversion by nonlinearity (B) Evaluation of quasi-crystal PBG	(E) Singlemode high power laser (E) Beam-controlled laser (B) High efficiency beam-controlled LED
	Buried Grating	Kyoto U. (APL75,316) U. Illinois (APL70.1119)	DF OX BN	GaAs InP	(E) Operation in 2-D DFB laser with singlemode and single polar.	(E) Singlemode high power laser (B) High efficiency LED
	Multilayer on Grating	Tohoku U. (EL35,1272)	BS	Si TiO$_2$	(G) Polar. selectivity (B) Index confine- ment waveguide (B) Functional device integration	(B) High performance polarizer (B) WDM circuit, dispersion conpensator
	VCSEL Array	U. Würzburg (PRL81,2582)	EP +DE	GaAs	(B) Discussion on coupled VCSEL modes and atom physics	(B) Mode control (B) Wavelength plate, polarizder
	Sphere Array	Nat'l Res. Inst. Metals (JAP87.7152) U. Tokyo (PRL82,4623)	MN	Poly -styrene etc.	(B) Discussion on coupled sphere modes and atom physics	(B,E) Mode control, Ultra-high Q, Large non -linearity
	Fiber	Georgia Inst. Tech. (EL37,no25) Los Alamos Nat'l Lab. (OL26.1158) NTT (OE9,676) Tech. U. Denmark (EL37,no5) U. Auckland (OL26,1356) U. Bath (SC285,1537) U. Sydney (OE7,88)	FB	SiO$_2$	(G) Evaluation of PBG (B) Low loss (B) Large core, singlemode (B) Dispersion control (B) Large nonlinearity, super-continuum light generation (B) Amplifier, lasing by Yb,Er dope (B) Raman scattering, 4 wave mixing (B) Band pass filter	(B,G) Ultralow loss, high reliability (B) Arbitrary spot size converter (B) Low loss dispersion conpensator (B) Introduction to nonlinear systems (B,G) Wide practical use (B,G) Application to POF (B,G) Application to other device design
	Metal Patch with Post	Hard Rock Lab. (MTT47,2059)	MF	Cu	(G) Evaluation of PBG (G) Evaluation of antenna return loss	(G) High performance compact antenna

Dimension	Structure	Organization (Publication[1]) (A,B,C — order)	Fabri-cation[2]	Material[3]	Activity[4]	Prospects, etc.[4]
2-D	Planar Metal Patch	UCLA (MTT47,2123) U. Michigan (MTT47,2099)	MF +WE	Cu	(G) Strip line (G) Strip reflector	(G) Low loss strip line (G) High performance strip filter
3-D	Woodpile	Bilkent U. (APL74,486) Caltech (NT398,51) Electro-Optics Div. (APL77,3221) Kyoto U. (APL75,905) Sandia Nat'l Lab. (NT394,951) U. Tokushima (APL74,786)	BN DP +PS +DE MPA FL	GaAs InP Si Polymer Glass	(G) Evaluation of PBG (G) Evaluation of PL w/wo defect (G) Transmission of short pulse	(G,E) Low threshold lasing (G) Large scale ultra high density functional photonic circuit (G) All optical switch (B) High speed and high efficiency LED (B) Negative index optical component (B,E) Delay line, disper compen -sator (B,E) Large nonlinear device
	Cubic Lattice	ERATO (APL79,1228) Inst. Microtech. Mainz (APL71,1441) Osaka U. (APL76,2668) Pennsylvania State U. (APL75,2533) Shinshu U. (APL70,2966)	EP +DE +WE XL FL	GaAs Polymer	(G) Evaluation of PBG	(B) Filter,dispersion conpensator
	Oblique Holes	Inst. Microtech. Mainz (APL71,1441) UCLA+Caltech (JVST14,4110) U. Oxford (NT404,53) U. Paris (APL77,2942)	DE XL IL FIB	GaAs Polymer	(G) Evaluation of PBG	(G,E) Low threshold lasing (B) High speed and high efficiency LED (G,B) Functional bulk devices
	Buried Dots	U. Michigan (APL75,1670)	EP +DF +OX	GaAs	(G) Evaluation of PBG	(G) Cavity, waveguide, etc. with defects
	VCSEL Array	UCLA (AO37,2074) Swiss Federal Inst. Tech. (NT407,880)	EP +DE	GaAs	(B) Nonlinearity in coupled modes	(B) Mode control in VCSEL
	Multilayer on Grating	NEC (PRB58,10096) Tohoku U. (EL34,1260) U. Michigan (APL78,3024)	BS EP +OX	Si GaAs	(B) Evaluation of bands (B) Strong dispersion	(B)WDM μ-filter (B) m-dispersion conpensator (B) Functional bulk devices
	Multilayer on Grating with Holes	NTT (APL77,4256)	BS +DE	Si	(G) Evaluation of PBG	(G) Platform of photonic circuit
	Stacked Slab	Harvard U. (APL77,2098) MIT (APL73,145)	MF?	Al	(G) Evaluation of PBG	(G) μ-wave filter

Dimension	Structure	Organization (Publication[1]) (A,B,C — order)	Fabrication[2]	Material[3]	Activity[4]	Prospects, etc.[4]
3-D	Opal	Clemson U. (APL75,1497) Clemson U. (OL25,1300) DERA (EL36,no16) Glasgow U. (JLT17,2121) Ioffe Phys.-Tech. Inst. (APL75,1057) Jilin U. (OSA17,476) KAIST (JAP90,2042) Kyushu U. (APL72,1957) MIT (APL67,2138) Madrid Mat. Inst. Science (APL71,1148) Nanyang Technological U. (APL76,3513) Nat'l Inst.Advanced Industrial Science and Tech. (APL80,192) Nat'l Inst. Mat. Science (JJAP40,L1226) Nat'l Res. Inst. Metals (JAP87,7152) NEC Res. Inst. (APL76,1627) NJ Inst. Tech. (APL78,1754) Osaka U. (APL75,932) Pennsylvania State U. (JAP88,406) Princeton U. (NT401,893) Strasbourg Mat. Inst. Phys. Chem. (JJAP37,L1527) U. Colorado (PRL86,4052) U. Moncton (APL78,52) U.S. Army Res. Lab. (NT393,445) USC (JLT17,1956) U. Sheffield (APL78,4094) U. Tokushima (JJAP37,L1527) Utrecht U. (APL80,49) Wayne State U. (PTL12,1647)	SO MN	PS Si Ge SiO_2 GaAs InP ZnS LC etc.	(G) Evaluation of PBG (B) Evaluation of PL (B) Introduction to pigments	(B) Functional bulk device (B) Color fixing agent
	Inverse Opal	Allied Signal Inc. (SC282,897) Chinese Academy of Sciences (APL80,1879) Iowa State U. (APL74,3933) NEC Res. Inst. (NT414,289) Osaka U. (APL73,3506) Rice U. (PRL83,300) U. Amsterdam (PRL83,2730) U. Amsterdam (SC281,802) UCSB (NT389,948) USC (JLT17,1956) U. Toronto (JAP90,5328) U. Toronto (JLT17,1931) U. Wuppertal (APL79,731)	IO	TiO_2 C Polymer LC etc.	(G) Evaluation of PBG (B) Evaluation of PL (B) Introduction to pigments	(B) Functional bulk device (B) Color fixing agent
	Spiral	U. Alberta (NL2,59) U. Toronto(NL2,59)	GLAD	Silica		(G) Cavity, waveguide, etc. with defects

3.3 2D PHOTONIC CRYSTALS MADE OF SEMICONDUCTORS

3.3.1 General features

2D crystals have characteristics intermediate between those of 1D and 3D structures in terms of their ease of fabrication and expected properties. The fabrication process is more complex for 2D crystals than 1D crystals since a film deposition technique can be used to produce 1D crystals, whereas it is necessary to use a patterning technique for 2D crystals. However, it is simpler to fabricate 2D crystals than 3D crystals, which require both film deposition and patterning techniques. As the present patterning technique is rapidly improving along with Si large-scale integration (LSI), many developments should be achieved by utilizing this technique for 2D crystals. Many applications of photonic crystals are directed towards the integration of photonic devices. Therefore, it is important for 2D crystals to evolve into the photonic and electronic circuits, which will require high-density photonic and electronic circuits. Furthermore, 2D crystals can be used in a multiple layer structure, whose each layer may possess different functionality, such as interconnections, electric circuits, photonic circuits, and so on.

On the other hand, expected properties of photonic crystals can be roughly divided into the following: (i) a photonic bandgap (PBG), (ii) anomalies in dispersion and anisotropy and (iii) a low group velocity. The first of these properties, a complete 3D PBG, cannot be formed in 2D crystals because the structure possesses a periodicity in the 2D plane that results in light propagation in the vertical direction. However, in a hole-arranged structure made into a high refractive index film (a semiconductor, for example) that is sandwiched between low refractive index mediums (air, for example), light is confined in the vertical direction by total internal reflection. Such a structure is called a photonic crystal slab and has recently been studied extensively because it generates a pseudo-PBG covering almost all of the 3D angles and is easily processed. Also, it has the most suitable structure for above-mentioned photonic and electronic integrated circuits. Properties (ii) and (iii) occur even in 1D crystals. The characteristics of 2D crystals include complex diffraction, a number of structural parameters, a variety of functions, and a degree of design freedom. In a 1D crystal, it is mainly the periodicity, along with the direction of light propagation, that generates the anisotropic effect. For 2D crystals, the periodicity in an orthogonal direction also affects the direction of light propagation, and changes the band number, the inclination and the curvature. This is the essential difference between 1D and 2D crystals. Regarding this point, the difference between 2D and 3D crystals is rather small.

3.3.2 Possible applications expected from photonic bands

Examples of photonic bands are shown in Fig. 3.3.1. When the structure length is infinite in the vertical direction and the wavevector k lies in the photonic crystal's plane of 2D periodicity, the band is separated into the transverse electric (TE) polarized band, where the electric field is parallel to the plane and the transverse magnetic (TM) polarized band, where the magnetic field is parallel to the plane, as indicated in Fig. 3.3.1(a). The precise definition of TE and TM differs between publications, but when k is normal to the periodic direction in a 1D crystal, these two bands are completely degenerate. In a 3D crystal, polarized waves cannot be separated. Thus, this separation of polarized waves is a unique characteristic of 2D crystals. Figure 3.3.1(b) shows the band of a photonic crystal slab. Here, the horizontal axis denotes horizontal components of k. When the structure is symmetrical in the vertical direction, some modes can be separated by the symmetric or anti-symmetric characteristics of the electromagnetic field in the vertical direction. Considering the main polarized waves, separated waves are sometimes termed TE-like and TM-like polarized waves. In a similar way to 3D crystals, polarized waves cannot be separated in an asymmetric structure of a 2D crystal. As seen in (a) the band obviously

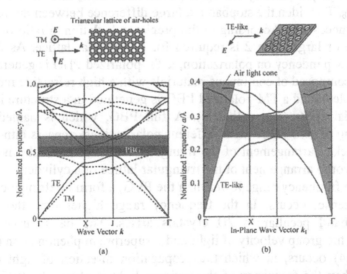

Figure 3.3.1 Band diagram of a circular hole 2D photonic crystal. (a) Band calculated by the plane wave expansion method for both polarized waves in a triangular lattice crystal, which is uniform in the vertical direction. (b) Band calculated by the FDTD method for a TE-like polarized wave in a photonic crystal slab. The dark gray horizontal band and the light gray band represent the full PBG for both polarized waves and the PBG for one polarized wave, respectively. Inclined gray zone in (b) identifies the lightcone, which indicates the light radiation condition.

differs by polarization and this has prompted investigation into an application technique for individual polarized waves. This strong polarization dependence can be actively utilized as a polarization selective filter.

A PBG that prohibits both transmission and generation of light is an extended version of the stopband found in multilayers and diffraction gratings. In 2D crystals this phenomenon produces a PBG, which generates a stopband against light propagating in an arbitrary direction in the 2D plane. In order to obtain such a PBG in a 2D crystal, the Brillouin zone, which represents the frequency characteristics and direction dependence, should ideally be circular. For this reason, a triangular lattice is often discussed, where the Brillouin zone becomes hexagonal. The length from the Brillouin zone center to the edge (Γ-X) differs by a factor of 0.86 compared to the length from the center to one vertex (Γ-J). This difference is 0.71 for a square lattice. The position of the stopband shifts at these ratios and is dependent on the light forwarding direction. It is necessary that the half width of this stopband exceeds the change of the stopband position. For a 1D crystal, the width of the stopband with respect to its center frequency is formularized as $\Delta\omega_{SB}/\omega_{SB} = (4/\pi)\sin^{-1}\{(n_a-n_b)/(n_a+n_b)\}$ using the refractive indices of the periodically arranged photonic atom n_a and the background medium n_b. To widen the stopband, a large difference between the refractive indices is necessary. According to the precise calculation, a ratio of n_a/n_b or its inverse of larger than 2 is required for a triangular lattice. As the PBG also has a dependency on polarization, a TE-polarized PBG is generated in a structure composed of continuous materials with a high refractive index (e.g. circular holes), and a TM-polarized PBG is generated in a structure in which the material splits (e.g. cylinders). A full PBG, which is named as the overlapping of two PBGs with different polarizations, appears either in the closely-packed arrangement of the triangular lattice of a circular hole or in the honeycomb arrangement of the triangular lattice of a cylinder.

In the frequency range lower than the PBG, a form birefringence, in the classical sense, occurs. In the frequency range higher than the PBG, a complex band peculiar to 2D crystals arises. The band curve's slope represents the group velocity of light and a superprism phenomenon (refer to Section 2.4) occurs, in which the propagation direction of light changes sharply when the frequency of the incident light is above the PBG in this frequency range. By utilizing this phenomenon, either a wavelength filter or a dispersion compensator can be formed. Although these frequency ranges do not necessarily require a difference in the refractive index between the mediums, a large difference in the refractive index is still effective in emphasizing the peculiarity of the dispersion and the anisotropy.

Inclination of the k vector in the direction vertical to the 2D plane narrows the PBG and a large inclination results in its extinction. This is the main shortcoming of these crystals compared with a 3D crystal PBG. The PBG of the photonic crystal slab becomes narrower than that of a simple 2D crystal because the propagating light in the slab contains k vector components in the vertical direction. This slab structure gives birth to a light radiation area called the lightcone, which is indicated as a gray zone in Fig. 3.3.1(b). The lightcone is simply determined by the equation $\omega/c > k/n_c$, where n_c is the refractive index of the clad at both vertical sides of the slab. In the area below the lightcone, light propagates with complete confinement around the slab and in the area above the lightcone light propagates with a radiation loss. In such a structure, a PBG common to both TE-like and TM-like polarized waves is obtained when the hole diameter exceeds 80% of the period. As a large hole diameter results in a decrease of the average refractive index of the slab, so the PBG overlaps on the lightcone by shifting to a higher frequency range, and the practically available PBG becomes rather narrow. If the periodicity is reduced, the PBG shifts to a lower frequency. It is difficult to fabricate such a slab because it requires narrow intervals between the holes. At present, many researchers are limiting the hole diameter to around 60% of the period and utilizing the PBG for TE-like polarized waves, as shown in Fig. 3.3.1(b).

3.3.3 Fabrication Techniques

Generally, lithography is used to fabricate patterns of 2D crystals. It is rather difficult to form a period of $0.2 - 0.5$ μm, which generates a PBG in the optical frequency range, by the usual photolithographic techniques; consequently, many researchers use electron beam (EB) lithography. However, as long as the objective wavelength is not in the ultraviolet range (i.e. a wavelength over 0.1 μm) either a stepper exposure or a multi-interference exposure will be used to implement the process. To determine the characteristics of a 2D crystal possessing infinite height it is necessary that the structure be sufficiently deeper than the period length of the crystal. At first it was thought that process techniques yielding a high aspect ratio were important and research into techniques such as dry etching, wet etching, anodic oxidation, and selective epitaxial growth has progressed. Recently this requirement has become less important because of the progress made in photonic crystal slab research.

Dry etching techniques include reactive ion etching (RIE) and reactive ion beam etching (RIBE). In the former method, an inductively coupled plasma (ICP) is further utilized to obtain high-density plasma. In the latter method a chemically assisted ion beam (CAIBE) is sometimes used. By using a rare gas to generate the ion beam and subjecting the substrate to a reactive gas flow, the flexibility of the etching condition is expanded. The degree of difficulty of etching varies according to the material. For example, ICP-RIE with a Freon family gas for Si etching realizes a high etching rate and a high aspect ratio. A III-V compound semiconductor is necessary for a light emitting device. Although a high aspect ratio is obtained for the GaAs system by using RIBE and CAIBE with a halogen family gas, the internal efficiency of light emission decreases in the photonic crystal because of its large surface recombination velocity. Although it is more difficult to process the InP system than the GaAs system, the recombination velocity of the former is lower than that of the GaAs system. The ideal etching characteristic for the InP system is being proven by RIE using the alternative change of a carbonized gas and oxygen, and by ICP-RIE using a chloride family gas. A fabricated example of a 2D crystal is shown in Fig. 3.3.2.[1] The GaN system is also a noteworthy semiconductor. Although this material has been studied from several aspects, the results lead us to expect a low recombination velocity because the internal efficiency is maintained even with the high dislocation density of the epitaxial film. Research into etching techniques yielding high aspect ratios has just begun, so notable progress in this field can be expected.

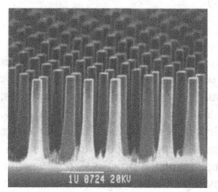 

Figure 3.3.2 Triangular lattice 2D crystals made of GaInAsP/InP with a cylindrical honeycomb arrangement (left) and a circular hole close-packed arrangement (right) fabricated by the chlorine gas ICP-RIE method.

The anodic wet etching technique and the anodic oxidation technique are already used for ultrahigh aspect ratio etching in which the hole reaches to the back of the wafer (refer to Section 3.5). The selective epitaxial growth technique is suitable for active devices. After covering the surface of the GaAs or GaN with an oxide and nitride possessing the desired periodical openings, metal organic vapor phase epitaxy (MOVPE) is carried out. Under special conditions, film growth in an upward direction occurs solely in the openings. Even if the original openings are circular, a cross-section of the growing part of the film shows a hexagonal shape, reflecting the crystal orientation. Inserting an active layer during the crystal growth process results in the fabrication of a light emitting device. However, the crystal does not always grow in the vertical direction and an inclined crystal facet can form as the aspect ratio increases. The shape of the cross-section is likely to change from a hexagon to a triangle or rhombus, which is undesirable.

3.3.4 Light emitting devices

The point defect laser shown in Fig. 3.3.3(a) is the device first suggested as an application for photonic crystals.[2] The laser has been highlighted mainly in order to obtain control of the spontaneous emission and thresholdless oscillation accompanying the microcavity and the introduction of non-uniform elemental active regions (which are called defects) in the uniform crystal. An alternative viewpoint has caused this device to be known as a distributed Bragg reflector (DBR) laser, because it resembles multi-dimensional Bragg's mirrors surrounding the defect cavity. A distributed feed back (DFB) laser that uses the whole defectless crystal as an active region, is shown in Fig. 3.3.3(b), and a light emitting diode (LED) with high efficiency where the light extraction efficiency increases, as depicted in Fig. 3.3.3(c), are covered in this subsection.

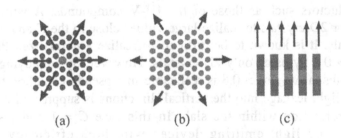

(a)   (b)   (c)

Figure 3.3.3 Concept of photonic crystals as emitting devices. (a) Defect laser, (b) DFB laser and (c) LED.

Present issues of defect laser development are the following three: (i) low threshold continuous wave operation at room temperature by photo-pumping, (ii) a large spontaneous emission factor C (or sometimes denoted as β. It is the coupling efficiency of spontaneous emission energy into the lasing mode) as high as $0.1 - 1$, and (iii) an enhancement of the spontaneous emission rate (Purcell effect) denoted by Purcell factor γ (or sometimes

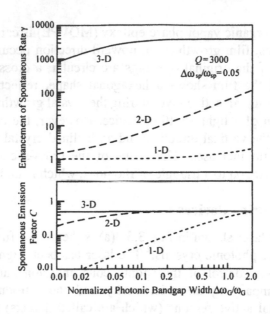

Figure 3.3.4 A simple estimation of spontaneous emission factor *C*, and the Purcell factor γ. As the polarization of the spontaneous emission is set as uniform, the upper limit of *C* is 0.5.

denoted as *F*). The theoretically induced *C* and γ values are shown in Fig. 3.3.4. The wide PBG is indispensable for strong enhancement of this effect.[3] As a large difference between the refractive indices of mediums is necessary for a wide PBG, the available materials are primarily direct–transition-type semiconductors such as those of the III-V compounds. A wider PBG is needed for 2D crystals to realize large values close to the *C* and γ values of 3D crystals. It is known to be difficult to realize a normalized PBG width $\Delta\omega_G/\omega_G > 0.2$ by using only 2D crystals. However, by employing photonic crystal slabs, a $\Delta\omega_G/\omega_G > 0.8$ is realizable in a pseudo condition because of the extra light leakage into the vertical directions is suppressed by the total internal reflection within the slab. In this case *C* = 1 and γ ~ 100 are available and light emitting devices with high efficiency and high spontaneous emission rate are expected to be realized. The successful lasing oscillation obtained with a photonic crystal slab defect laser in 1999 generated much interest. However, the surface recombination due to the direct processing of the active layers is a serious problem with this laser; it is desirable to obtain a small surface recombination velocity. Since 1995, the InP system has been studied as a suitable material for the suppression of the surface recombination of photonic crystal light emitters,[2,4] and has become a mainstay of the research efforts including the first demonstration of the defect laser. However, in the photonic crystal process, the ratio of exposed surface area to volume of active area is still excessively large. The surface recombination rate is obtained as the product of the typical surface recombination velocity of an InP system just after etching (e.g. 1.5×10^4 cm/s) and the typical exposed area to active volume ratio of the photonic crystal. It is larger than the standard radiative recombination rate of InP systems and the internal efficiency of light emission is as small as 20 – 30%. Accordingly, the lasing threshold so far reported is incomparably larger than those of other types of microlaser such as the microdisk. Suppression of the surface recombination by various treatments after the etching process was

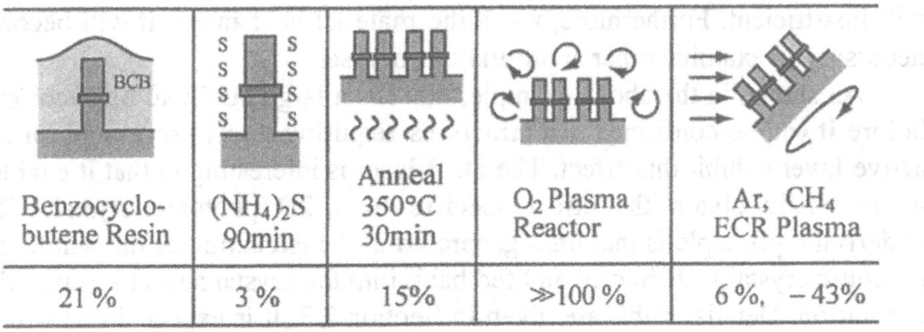

Benzocyclo-butene Resin	$(NH_4)_2S$ 90min	Anneal 350°C 30min	O_2 Plasma Reactor	Ar, CH_4 ECR Plasma
21 %	3 %	15%	≫100 %	6 %, −43%

Figure 3.3.5 Decreasing ratios of surface recombination velocity corresponding to various surface treatments for fine cylindrical photonic crystals of a GaInAsP/InP semiconductor.

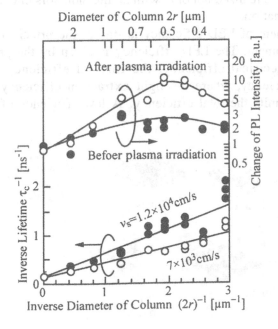

Figure 3.3.6 Cylinder diameter dependency of emission lifetime and emission intensity before and after methane plasma irradiation.

studied for a GaInAsP/InP 2D crystal.[5] As shown in Figs. 3.3.5 and 3.3.6, the surface recombination velocity decreased to 7×10^3 cm/s after irradiation of CH_4 plasma. Shallow implantation of carbon atoms from the surface to a depth of over 10 nm was observed by second ion mass spectroscopy (SIMS) analysis. The decrease in the surface recombination velocity is considered to be due to electrical insulation in the vicinity of the exposed surface area, where the carbon atoms form the deep doping levels of the GaInAsP active layer. However, as a surface recombination velocity of 1×10^3 cm/s is necessary to obtain the targeted high efficiency, the amount of suppression is still insufficient. Furthermore, when the material is changed, it will become necessary to examine other appropriate processes.

As shown in the above example, there is a large hurdle to be overcome before it can be confirmed that structures requiring direct processing on the active layer exhibit this effect. The DFB laser is interesting in that it enables us to easily obtain the effect specific to a 2D photonic crystal. The underlying principle is that light generated at the excitation of the whole 2D photonic crystal is diffracted and fed back into the crystal so as to trigger the oscillation. Details of this are given in Section 2.3. It is expected that it will be feasible to produce a high intensity laser with a single longitudinal and lateral mode and a large surface area. As this laser will not rely on a PBG, a large difference in the refractive index is not necessary. Thus, this lasing oscillation can be realized not only with semiconductors but also with a glass and dye combination.

Enhancement of LED efficiency is one of the principal themes of 2D crystal development. The LED efficiency is given by the product of three factors: the electrical efficiency, the internal efficiency, and the light extraction efficiency, where the light extraction efficiency is the largest element that limits the total efficiency. A low efficiency of 5% or less is

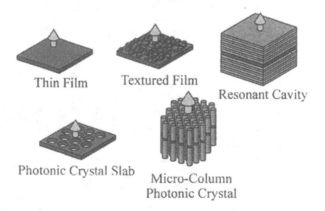

Figure 3.3.7 Various fine LED structures for increasing the light extraction efficiency.

mainly due to the total internal reflection at the boundary, in spite of the use of a curved mold with a cheap, transparent epoxy resin. Although a high efficiency (over 50%) can be obtained using a substrate made with special processing features, it is difficult to introduce a complicated process to the production of LEDs, where the low cost requirement is greater than for lasers. The possibility of increasing the efficiency of photonic crystal LEDs by using a semiconductor slab has been discussed for some time. As shown in Fig. 3.3.7, the principles of this discussion are as follows: (i) a decreasing effect in the modal refractive index of the thin film and the photon recycling effect in which light is reabsorbed in the active layer and regenerated as a

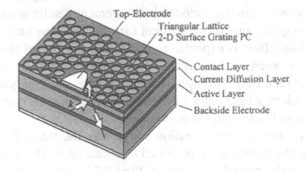

Figure 3.3.8 A surface diffraction grating type photonic crystal LED.

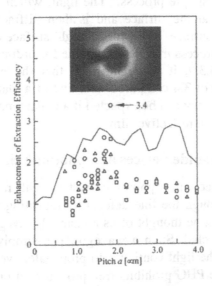

Figure 3.3.9 The increase of LED efficiency regarding the lattice pitch of a surface diffraction grating type photonic crystal LED. The solid line curve represents the calculated results, whilst plots show the experimental results. The photograph is a near-field pattern of light emission.

carrier; (ii) the extracting effect, where light reaching the face at a certain angle escapes from the total internal reflection at one side via a textured film surface after repetitive reflection; and (iii) the effect of the changing direction of light radiation, brought about by a film vertically sandwiched between reflectors of a multilayered semiconductor, on the light extracting effect. An efficiency of about 10% is realized from (iii), a process for which the electrode is easily made. However, its application is limited to near-infrared materials at present because multilayer fabrication requires a high degree of epitaxial technique. As shown in Fig. 3.3.1, photonic crystal LEDs realize a high efficiency with positive utilization of the lightcone. Elimination of the wavevector that moves forward along the plane of the structural wavevector in the periodical arrangement results in a change of the light forwarding direction to the vertical direction and the extraction of light into the air. This effect is expected to occur in a 2D crystal that includes an emission layer such as a circular hole arrangement or fine cylinder arrangement on the film. Through photopumping experiments using crystals with fine cylinders, a light extraction effect over 20 times larger (about 40%) than that of an ordinary wafer was evaluated.[1]

However, surface recombination suppresses the total efficiency of the LED because the active layer is directly processed in the case of the two types of photonic crystals shown in Fig. 3.3.7. The surface diffraction grating-type photonic crystal, shown in Fig. 3.3.8, avoids this problem and can be realized by a simple process.[6] The light, which is radiated from the active layer, arrives at the surface and is then diffracted into a direction perpendicular to the surface. Since just a slight surface etching is necessary for fabrication, this process may be acceptable for ordinary LED production. As shown in Fig. 3.3.9, it was confirmed that the maximum efficiency increased by a factor of 3.4 compared with that of a plane-type LED. A 10-fold increase in efficiency can be expected if a plastic mold is used and if the substrate is backed by a reflective film.

3.3.5 Optical waveguide devices and photonic circuits

By inducing line defects in a photonic crystal, light in a certain frequency range propagates through the line defects by repeating the PBG reflection. These line defects can be thought of as a channel waveguide surrounded by the photonic crystal clad. Such a waveguide is simply called a *photonic crystal waveguide*. The light continues to propagate, even in arbitrarily bent channels, because the PBG prohibits free propagation in the direction along the PBG plane. That is, the light either propagates forward or reflects backward when it reaches a sudden bend. However, some frequencies of light propagate in the bend with a probability of nearly 100% based on the

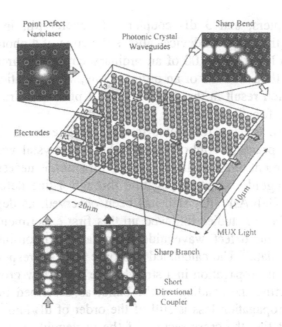

Figure 3.3.10 Schematic of a dense photonic crystal circuit composed of a nano-laser, an optical waveguide, and waveguide-type optical devices. The power distribution of propagating light was calculated by the scattering matrix method.

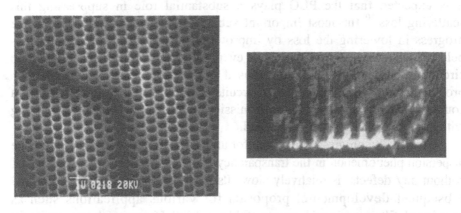

Figure 3.3.11 A circular hole photonic crystal fabricated on a GaInAsP slab (a), and the near-field pattern of light propagation for a TE polarized wave with a wavelength of 1.53 μm (b). The reversed bend in propagation direction between (a) and (b) occurs because the observation was made from the reverse side of the photonic crystal.

same theory as the Fabry-Perot resonator. Furthermore, this frequency range can be expanded by slightly lessening the amount of curvature. It is easy to introduce a 90° bend into a square lattice crystal and a 60° or 120° bend into a triangular lattice crystal. Thus, it will be possible to produce, sharp branch, directional couplers, and 3 dB couplers of several μm in length.[7] The combination of these optical devices allows the edge of a photonic circuit to be shortened to less than 1/100 of an ordinary circuit. Its area can then be reduced to less than 1/10000 of an ordinary one, as schematically shown in Fig. 3.3.10, which result in the development of photonic circuits with very high density and functionality.[8]

It is only in the last few years that the real research for testing the theories on the practical applications of photonic crystal waveguides has started. Confinement of the light in a vertical direction is necessary to form a 2D crystal waveguide. Light propagation through a line defect waveguide fabricated in a GaInAsP/SiO$_2$ slab has been observed, as depicted in Fig. 3.3.11.[9] A number of studies followed up this first experiment by trying to form a single line defect waveguide using a silicon-on-insulator (SOI) substrate as the slab.[10] The calculated photonic band corresponds well with the observed light propagation in a straight line and a low group velocity is observed, reflecting the bending of the band, as described in Section 3.4. However, the propagation loss is still of the order of dB/mm.[11] Taking into account the fact that the cross-section of the waveguide is much less than one square μm and that the difference between the refractive indices of the mediums is 45% or more, it is reasonable to assume that this large loss results from a large scattering loss, due to incomplete fabrication. Although it is expected that the PBG plays a substantial role in suppressing this scattering loss,[10] the most important subject in the near future is to make progress in lowering the loss by improving the process. A maximum loss below 1 dB/cm would be required, even for use in high density photonic circuits. If a line defect waveguide based on the photonic crystal slab is to be produced and used in high density circuits, a small bend, a branch, a cross, a coupler with low loss, a wide transmission band, and an efficient coupling with external light must be developed.

Conversely, progress in research for utilization of the *superprism*, a unique dispersion phenomenon in the transparency band of a uniform photonic crystal without any defects, is relatively slow. Its first experimental observation and subsequent developmental proposals for various applications such as wavelength filters occurred between 1998 and 1999. However, there were many uncertainties relating to factors such as wavelength resolution, the number of resolved points that it possessed, and the required conditions for incident angle, beam spotsize and the prism length. These problems were addressed by a theoretical investigation based on the dispersion surface analysis.[12] As shown in Fig. 3.3.12, a parameter indicating the degree of resolution of the superprism against a realistic incident light beam with a finite width is mapped over the Brillouin zone. It indicates that a high resolution is not obtained by an abruptly changing dispersion characteristic but by a flat characteristic that changes its

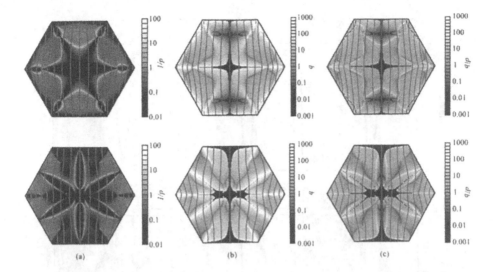

Figure 3.3.12 Calculated parameters showing light beam collimation (a), wavelength sensitivity (b) and wavelength resolution (c) at fiber communication wavelengths in a 2D photonic crystal superprism of hole arrangement in a triangular lattice. The refractive index of the background medium is 3.065, and the ratio of hole diameter to the lattice pitch is 0.624. Gray vertical curves denote equi-incident-angles with an interval of 5°.

angle with wavelength. A high resolution can be obtained in fiber communication wavelength range by setting the incident angle so that a high resolution parameter region lies along an equi-incident-angle curve.

In addition to the superprism, a large increase of nonlinearity due to the low group velocity is also expected for uniform photonic crystals. The response for a 2D crystal with third order nonlinearity was calculated by means of the FDTD method. The results indicate that low loss and a large nonlinearity can be expected at the second Γ point of the photonic band, as shown in Fig. 3.3.13. A nonlinear response showing an abrupt saturation with use as an optical limiter can also be anticipated.[13] The loss, accompanied with the reflection of incident light into the photonic crystal from the outer area, is a serious problem with these devices. A moderate change in the structure at the boundary surface is effective in avoiding this reflection loss. Calculations have shown that it is nearly extinguished within a range of 10% of the center frequency in the structure presented in Fig. 3.3.14.[14]

References

1. T. Baba, K. Inoshita, H. Tanaka and J. Yonekura, J. Lightwave Technol. **17**, 2113 (1999).
2. T. Baba, and T. Matsuzaki, Electron. Lett. **31**, 1776 (1995); Jpn. J. Appl. Phys. **35**, 1348 (1996).
3. T. Baba, Fundamentals and Applications of Photonic Crystals (Ed. S. Kawakami), CMC Publishers, 185 (2002).

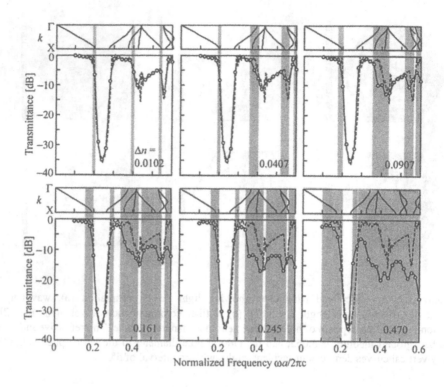

Figure 3.3.13 Calculated result of transmission spectra for an incidental plane wave from the Γ–X direction of a circular hole arranged triangular lattice 2D crystal. The dotted line is assumed to be a linear medium. Solid line and plots are assumed to be a third order nonlinear medium. The Δ*n* represents the dependence of the refractive index change of the nonlinear medium on the intensity of the incidental light.

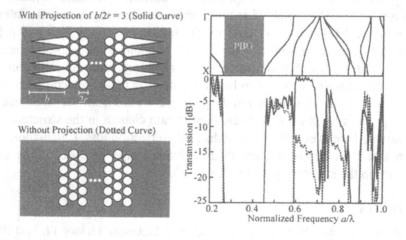

Figure 3.3.14 A transmission efficiency calculation made at a frequency range higher than the PBG of a 2D crystal. The solid and the dotted lines correspond to a projection-type interface structure and an ordinary crystal, respectively.

4. T. Baba, IEEE J. Sel. Top. Quantum Electron. **3**, 808 (1997).
5. H. Ichikawa, K. Inoshita, and T. Baba, Appl. Phys. Lett. **78**, 2119 (2001).
6. T. Baba, and H. Ichikawa, Proc. Optoelectronic and Commun. Conf., 9C2-1 (2002).
7. J. Yonekura, M. Ikeda, and T. Baba, J. Lightwave Technol. **17**, 1500 (1999).
8. T. Baba, Inst. Electrical, Information, and Commun. Eng. **81**, 1067 (1998, in Japanese).
9. T. Baba, N. Fukaya, and J. Yonekura, Electron. Lett. **27**, 654 (1999).
10. T. Baba, A. Motegi, Iwai, N. Fukaya, Y. Watanabe, and A. Sakai, IEEE J. Quantum Electron. **38**, 743 (2002).
11. T. Baba, N. Fukaya, and A. Motegi, Electron. Lett. **37**, 761 (2001).
12. T. Baba, and T. Matsumoto, Appl. Phys. Lett. **81**, 2325 (2002).
13. T. Baba, and T. Iwai, Jpn. J. Appl. Phys. **42**, (2003).
14. T. Baba, and D. Ohsaki, Jpn. J. Appl. Phys. **40**, 5920 (2001).

(by T. Baba)

3.4 FABRICATION OF 2D PHOTONIC CRYSTALS COMPOSED OF SILICON-BASED MATERIALS AND ORGANIC MATERIALS

3.4.1 SOI photonic crystal slab

Photonic crystals have many unique and attractive characteristics,[1] which are quite different from all other materials. Since many of these characteristics are suitable for integration of photonic devices, photonic crystals can be a key platform for the realization of ultrasmall large scale photonic integrated circuits in the future. As 1D crystals lack the degree of freedom necessary for circuit constitution, photonic crystals with two-dimensions or more are required to form a photonic circuit. Since an ordinary circuit consists of a 2D plane and 2D crystals are far easier to fabricate than 3D crystals, researchers at many different institutions worldwide are presently engaged in developing 2D crystals for photonic integrated circuit applications.[2]

Among the various types of 2D crystals, the photonic crystal slab structure using a silicon-on-insulator (SOI) substrate, has some advantages for future application as a photonic circuit. The SOI substrate, which is a three layered structure like $Si/SiO_2/Si$, as shown in Fig. 3.4.1(a), was recently developed for use with the most advanced LSI and was found to be very promising for photonic crystal fabrication because of its high uniformity and high optical qualities. Silicon has not been used for ordinary photonic devices because fine control of the refractive index is difficult and also it is an indirect-transition-type semiconductor. However, as a material for photonic crystals, it has several advantageous characteristics, such as a large refractive index (about 3.5) and a remarkably low absorption loss in the wavelength range of optical communications. In addition, we can make use of many mature process techniques, which enable fabrication of hetero-structures that have a large index contrast, such as Si/SiO_2. Most importantly, ultrafine process techniques developed for electronic integrated circuits can now be utilized for photonic crystals, where nanometer-scale accuracy is required.

If we have sufficiently high-grade lithography technology and the ability to transfer patterns onto high index materials, it is not difficult to realize 2D photonic crystals with a large PBG. The most important question relating to the production of useful 2D photonic crystals is how to accomplish the vertical confinement of lightwaves within the 2D plane of the photonic crystal. The structure known as a photonic crystal slab has been extensively studied and has been found to meet the above requirements.[3] This structure consists of a sandwich of two types of material. A core made from a high refractive index material like a semiconductor is sandwiched between cladding layers of a low refractive index material, such as air or an oxide. The SOI substrate itself is intrinsically a slab waveguide structure with strong ability to confine light. The fabrication of a photonic crystal on the upper Si layer leads to a photonic crystal slab. Besides the advantages of the Si material listed above, the use of an SOI type photonic crystal slab makes it relatively easy to fabricate a high quality photonic crystal, compared with other fabrication techniques. This is because the SOI substrate is of a very high quality and a very high accuracy can be obtained (SOI wafers are now being used for the most advanced microprocessor units, in which the required homogeneity of the thickness is just a few nm). This is very important since, generally, uniformity and absorption loss of the original slab waveguides have considerably affected the characteristics of photonic crystals.

Figure 3.4.1(b) shows a schematic picture of an SOI type photonic crystal slab. Although a detailed explanation may be unnecessary, the process used to fabricate the photonic crystal is remarkably simple; it occurs in one step, a combination of a lithography technique, e.g. EB lithography, and a dry etching technique, e.g. electron-cyclotron-resonance (ECR) plasma etching, both of which are commonly used processes for Si.

3.4.2 Experimental confirmation of PBG

A scanning electron microscopy photograph of a 2D SOI photonic crystal slab is shown in Fig. 3.4.1(c).[4,5] An airhole pattern with a triangular lattice is formed on the SOI substrate by the use of EB lithography and ECR plasma etching.[6] The thickness of the Si layer and SiO_2 layer are 200 nm and 3 μm, respectively. The light is strongly confined to the photonic crystal slab between the upper air cladding layer and the lower SiO$_2$ (n = 1.46) cladding layer. The thickness of the Si layer is determined to be about one half of a wavelength of the light that will propagate through the waveguide and satisfies the single mode condition for slab waveguides.

The presence of a PBG in the 2D plane of a 2D photonic crystal is a major premise for the realization of a photonic circuit. The measurement of transmittance in the 2D plane is the most direct verification of the formation of a PBG. The transmission spectra along two crystal axes (Γ-M and Γ-K directions) of the SOI photonic crystal slab were measured and are shown in Fig. 3.4.1(d). The obvious 2D PBG is generated in the wavelength range of 1.31– 1.58 μm. This result corresponds well with the theoretical calculation using the 3D FDTD method. As the obtained PBG covers almost the entire wavelength range used for fiber communication, the introduction of line defects or point defects into this SOI photonic crystal slab will enable the use of various optical components.

The structure shown in Fig. 3.4.1(b) uses an oxide layer as the lower cladding. Because it is one of the basic structures, the airbridge-type slab, whose upper and lower cladding are replaced with an air layer, has also been widely studied.[7-10] However, oxide-type slabs are thought to be more favorable to practical applications, especially for large-scale integration

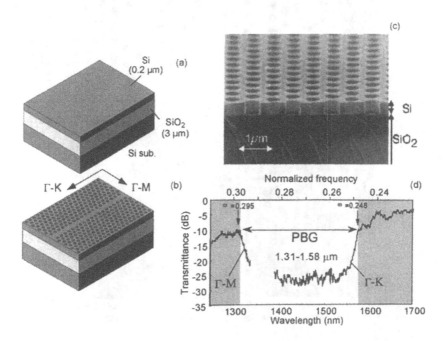

Figure 3.4.1 SOI photonic crystal slab. (a) SOI substrate, (b) typical model of an SOI photonic crystal slab possessing a line defect, (c) electron microscope image and (d) transmission spectra.

which needs larger photonic crystals. However, in the oxide-type clad, it has been pointed out that a true PBG is missing as a result of coupling between TM polarized waves and TE polarized waves due to the lack of symmetry in the vertical direction. A numerical investigation of this TE/TM polarization mixing effect, in a vertically asymmetric photonic crystal slab on SiO_2 cladding, shows that the mixing effect is very small and that the extinction ratio of polarization is over 35 dB, as long as the PBG for the lowest frequency TE mode is used. In practice, therefore, oxide-type photonic crystal slabs do indeed have meaningful PBGs. Of course, this problem can be completely eliminated by the use of an oxide or a polymer as an upper cladding material.[11]

3.4.3 Line defect waveguides in SOI photonic crystals slabs[12]

Since the previous results have shown that SOI photonic crystal slabs have an PBG, functional defects that operate in this PBG can now be

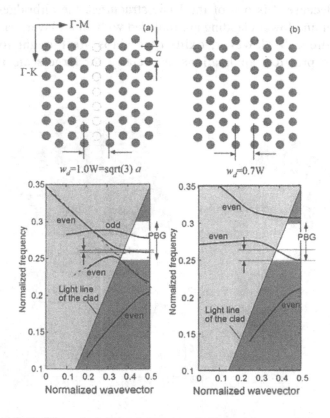

Figure 3.4.2 Theoretical dispersion curve of a linear defect waveguide. (a) $W_d = 1.0W$, and (b) $W_d = 0.7W$, where W means the channel width of the waveguide by missing one line of airholes.

considered. It is well known that the introduction of a point or line defect into a photonic crystal leads to functional operation as either an optical resonator or a waveguide, respectively. Here, a line defect waveguide[13,14] is investigated. A defect surrounded by a material with a PBG theoretically causes no loss. However, for a photonic crystal slab, it is necessary to take the radiation loss into account because the confinement in the vertical direction is not by the PBG but by total internal reflection based on the refractive index difference.[15] Figure 3.3.2(a) shows the theoretically calculated result for the dispersion curve of a single line defect waveguide of an SOI photonic crystal slab with a triangular lattice, in which a single row of holes is removed. The dispersion curve of line defect waveguides is considerably different to that of conventional waveguides, such as optical fibers, which confine light by a difference of refractive index. This dispersion curve looks complicated but can be understood qualitatively as a combination of gap-guided (GG: S-shape) and index-guided (IG: linear) modes. The significant guided modes have the characteristics of (i) existing in the PBG and (ii) having a long propagation distance. A mode above the light line of the cladding material becomes leaky and cannot propagate over long distances; true guided modes in the crystal slab exist under the cladding light line. These two points show that the true guided mode of a single line defect waveguide on an SOI photonic crystal is present in the field indicated by the arrow mark. As shown in the curve, there is no gradient in PBG, hence, the group velocity is very slow and the waveguide range is very narrow. Therefore, this seems to be a problem for a practical waveguide application using this type of line defect. The slow group velocity in the PBG is attributable to the anti-crossing of a GG mode and an IG mode that have an opposite mode curvature sign to each other in this frequency region. A mode with the desired group velocity in the PBG region can be realized by means of structural tuning of the line defect waveguide.[16] The simplest of these methods is to reduce the width of the single line defect waveguide. The dispersion curve of a line defect with a width reduction of 30% is shown in Fig. 3.4.2(b). A mode with a large group velocity appears in the PBG, which is different from the anti-crossing case.

Width-reduced line defect waveguides was fabricated and their transmission spectra were measured, as shown in Fig. 3.4.3. In this figure, one is a normal line defect waveguide and the other is a width-reduced line defect waveguide. A clear light transmission in the PBG is observed for the waveguide with reduced width, whilst no transmission is observed for the waveguide of normal width. Waveguides of various widths was fabricated and the spectra were successively measured so as to compare them with the theoretically calculated results. This work quantitatively confirmed that the theoretically designed modes correspond to the transmission mode observed

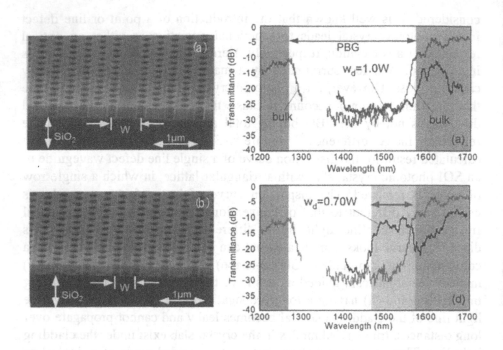

Figure 3.4.3 Transmission characteristics of a fabricated linear defect waveguide. Electron microscope image of (a) $W_d = 1.0W$ and (b) $W_d = 0.7W$, and the measured results of the transmission spectra of each waveguide.

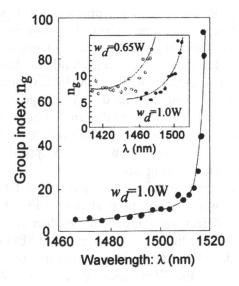

Figure 3.4.4 Measured wavelength dependency of the group refractive index.

in the PBG.[17] The propagation loss of guided waves was also evaluated by measuring the transmission through the fabricated samples with different waveguide lengths, in a similar way to the cutback method. The line defect waveguide with a width of $0.7W$ gave a loss of 6 dB/mm.[18] This is the smallest propagation loss value reported for a single mode photonic crystal waveguide; when normalized to 100 μm, it shows a loss of under 1 dB. It can be said that this loss value is already practical for an ultrasmall photonic circuit with a size of about 100 μm, though further reduction of the loss may be necessary for useful applications.

These results show that an adequate structural design raises the possibility of producing an efficient waveguide with a good transmission property in an SOI photonic crystal slab. It is important to mention that a tuned waveguide completely satisfies the single mode condition in the PBG, even if the leaky mode on the light line is taken into account. The effectiveness of the structural tuning is indicated in this point, too. Although tuning by reducing the width is introduced here, other methods of structural tuning are possible and recently good light transmission with phase-shifted hole-array type line defects were observed.[19] Recently, the dispersion of the group velocity in a photonic crystal slab single line defect waveguide was also measured by directly utilizing its own Fabry-Perot interference pattern.[20] The measured group velocity refraction index shows an extremely large group dispersion and group index (corresponding to a very small group velocity), as is shown in Fig. 3.4.4. The group index reaches a maximum of 90, which indicates a reduction in the energy propagation velocity of light down to 1/90 of the velocity in air. This slow group velocity and large group dispersion is expected to be effective for research into light-matter interactions such as optical nonlinearity.

3.4.4 Hybridization with 3D structures [21]

Although a good 2D photonic crystal can be made from an SOI type photonic crystal through a facile process, small but finite leakage of the light to the vertical direction, in a microcavity for example, cannot be suppressed completely because the vertical confinement depends on the classical total internal reflection, which is due to the refractive index difference. Therefore, it may become necessary to build up the vertical confinement by using a photonic crystal structure. With such a possibility in mind, a hybrid-type photonic crystal is being fabricating, using a $Si/SiO_2/air$ hybrid structure, by substituting the upper and lower layers of the 2D photonic crystal with 3D photonic crystal. The conceptual structure and the electron microscopic photograph of the fabricated hybrid type photonic crystal are shown in Fig. 3.4.5. A 2D Si/air photonic crystal formed using a lithography-and-etching

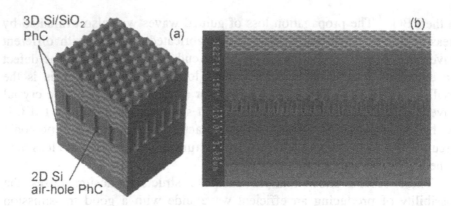

3D Si/SiO$_2$ PhC

(a)

(b)

2D Si air-hole PhC

Figure 3.4.5 (a) Schematic of a 2D and 3D hybrid-type Si/SiO$_2$/air photonic crystal and (b) an electron microscopic image.

technique is stacked on the 3D Si/SiO$_2$ photonic crystal made by the autocloning deposition technique, (refer to Section 3.7) after which, the next 3D Si/SiO$_2$ crystal is stacked by the autocloning deposition. As such a 3D crystal lacks a complete PBG, the confinement is not perfect, but it is expected that this type of 3D crystal may be more advantageous than the SOI photonic crystal in some cases. Also, the fabrication of a 3D crystal structure with a theoretical complete PBG[22] is formed by the combination of autocloning deposition and the lithography-and-etching technique, which is expected to be a perfect cladding layer for 2D photonic crystal slabs.

3.4.5 Organic photonic crystals

Though only the passive characteristics of photonic crystals have been mentioned in the previous subsections, it is important to incorporate some active functions (such as emission, switching etc.) into the discussion of photonic crystals, if one is to consider potential applications. With this in mind, the combination of a passive photonic crystal with a functional organic material is being studied. Organic materials have the following advantages: (i) various functional organic materials can be added to the photonic crystal platform, (ii) special processing specific to organic materials can be applied, (iii) organic materials are relatively resistant to processing damage, and (iv) organic materials are compatible with photonic crystals. An attractive final target is the fusion of Si-based photonic crystals and organic materials.

A special technique was developed, which enables us to selectively inject a functional organic material into a specially targeted single hole using the capillary phenomena.[23] The technique starts with very high-aspect-ratio 2D photonic crystals (alumina nano-hole array), which are fabricated by nano-printing lithography and an anodic-oxidation process. Then a through hole is formed at a targeted lattice point by means of a focused ion beam (FIB), from the back of the alumina nano-hole array. A polymer can be selectively injected into the targeted hole from the reverse of the sample as a result of the capillary force effect. The light emission is observed from the single hole where polychiophen (a light emissive polymer) is injected. Although the original 2D photonic crystal has no defects, the injected part of the crystal becomes a point defect cavity. In fact, the observed emission wavelength coincides with the point defect level of this crystal. The sharpness of the emission line is attributed to the strong confinement within the point defect cavity. The capability to dope various functional organic materials at arbitrary points without changing the original PBG of the photonic crystal is the key benefit of this process.

It is possible to fabricate organic 2D photonic crystals by the deposition or spin-coating of organic thin film layers on 2D periodic hole patterns on SiO_2.[24] This method can be used to produce organic photonic crystal lasers. Firstly, periodic holes are made in the SiO_2 film on a Si substrate by the lithography-and-etching technique. After this, the organic material is doped with a laser dye, (DCM doped Alq_3) the refractive index of which is higher than that of SiO_2. The dye is then evaporated to build up a 2D organic photonic crystal. Clear lasing action is observed by pulsed optical pumping of a N_2 laser. By changing the lattice period, various lasing actions due to different PBG edges are also observed, which reveal that the lasing feedback mechanism depends upon the type of PBG.

A photonic crystal laser structure almost identical to that mentioned above can be fabricated by a simple nano-printing technique, whilst the above structure is formed by time-consuming lithography-and-etching techniques.[25] First, a mold with a periodic hole pattern is prepared, which is a SiC substrate patterned by the EB lithography and an etching technique. Another substrate is prepared, on which a non-patterned thin film layer of an organic material (DCM doped Alq_3) is evaporated. Then, the periodic structure is transcribed onto the organic film simply by pressing the mold onto the second substrate. The same mold can be used many times. Lasing oscillation is observed for the sample fabricated by this nano-printing technique.

References

1. J. D. Joannopoulos, R. D. Meade, and J. N. Winn, Photonic Crystals, Princeton University Press (1995).
2. C. M. Soukoulis, Ed., Photonic Crystals and Light Localization in the 21st Century, Kluwer Academic (2001).

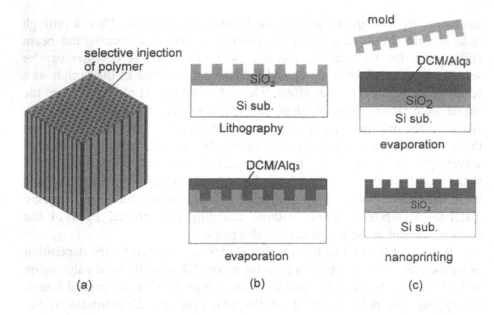

Figure 3.4.6 Method to fabricate photonic crystals using an organic material. (a) Fabrication of a high-aspect-ratio 2D photonic crystal with polymer injection, (b) dye evaporation on a patterned substrate, and (c) nano-printing technique.

3. S. Johnson, S. Fan, P. R. Villeneuve, and J. D. Joannopoulos, Phys. Rev. B **60**, 5751 (1999).
4. I. Yokohama, M. Notomi, A. Shinya, C. Takahashi, and T. Tamamura, Proc. Optoelectronic and Commun. Conf., 11B2-4 (2000).
5. A. Shinya, M. Notomi, I. Yokohama, J. Takahashi, C. Takahashi, and T. Tamamura, Opt. Quantum Electron. **34**, 113 (2002).
6. C. Takahashi, J. Takahashi, M. Notomi, and I. Yokohama, Proc. Mat. Res. Soc. **637**, E1.8 (2000).
7. N. Kawai, K. Inoue, N. Carlsson, N. Ikeda, Y. Sugimito, K. Asakawa, and T. Takemori, Phys. Rev. Lett. **86**, 2289 (2001).
8. M. Loncar, D. Nedeljkovic, T. Doll, J. Vuckovic, A. Scherer, and T. P. Pearsall, Appl. Phys. Lett. **77**, 1937 (2000).
9. O. Painter, R. K. Lee, A. Yariv, A. Scherer, J. D. O'Brien, P. D. Dapkus, and I. Kim, Science **284**, 1819 (1999).
10. S. Noda, A. Chutinan, and M. Imada, Nature **407**, 608 (2000).
11. A. Shinya, M. Notomi, and I. Yokohama, Proc. Integrated Photon. Res. Conf., (2000).
12. M. Notomi, A. Shinya, K. Yamada, J. Takahashi, C. Takahashi, and I. Yokohama, IEEE J. Quantum Electron. **38**, 736 (2002).
13. T. Baba, N. Fukaya, and J. Yonekura, Electron. Lett. **35**, 654 (1999).
14. M. Tokushima, H. Kosaka, A. Tomita, and H. Yamada, Appl. Phys. Lett. **76**, 952 (2000).
15. A. Chutinan and S. Noda, Phys. Rev. B **62**, 4488 (2000).
16. K. Yamada, H. Morita, A. Shinya, and M. Notomi, Opt. Commun. **198**, 395 (2001).
17. M. Notomi, A. Shinya, K. Yamada, J. Takahashi, C. Takahashi, and I. Yokohama, Electron. Lett. **37**, 293 (2001).

18. M. Notomi, A. Shinya, E. Kuramochi, I. Yokohama, C. Takahashi, K. Yamada, J. Takahashi, T. Kawashima, and S. Kawakami, IEICE Trans. Electron. E85-C, 1025 (2002).

19. K. Yamada, M. Notomi, A. Shinya, C. Takahashi, J. Takahashi, and H. Morita, Electron. Lett. 38, 74 (2002).

20. M. Notomi, K. Yamada, A. Shinya, J. Takahashi, C. Takahashi, and I. Yokohama, Phys. Rev. Lett. 87, 253902 (2001).

21. E. Kuramochi, M. Notomi, T. Kawashima, J. Takahashi, T. Takahashi, T. Tamamura, and S. Kawakami, Opt. Quantum Electron. 34, 53 (2002).

22. M. Notomi, T. Tamamura, T. Kawashima, and S. Kawakami, Appl. Phys. Lett. 77, 4256 (2000).

23. A. Yokoo, M. Notomi, H. Suzuki, M. Nakao, T. Tamamura, and H. Masuda, IEEE J. Quantum Electron. 38 938 (2003).

24. M. Notomi, H. Suzuki, and T. Tamamura, Appl. Phys. Lett. 78, 1325 (2001).

25. A. Yokoo and M. Notomi, J. Spectroscopical Soc. Jpn. (Bunkou-Kenkyuu) 51, 171 (2002, in Japanese).

(by M. Notomi)

3.5 HIGH ASPECT RATIO 2D PHOTONIC DIELECTRIC CRYSTALS

3.5.1 Fabrication of high aspect ratio

An important question in fabrication of 2D photonic crystals is how to achieve a high level of accuracy when forming a structure with a high aspect ratio in the dielectric. Generally, the most popular fabrication technique for producing 2D photonic crystals is a hole array or a pillar array structure based on a semiconductor using a dry-etching technique. However, it is difficult to obtain a high aspect ratio structure in 2D photonic crystals by using the above mentioned techniques. Conversely, there is a relatively simple process for fabricating a 2D photonic crystal based on high aspect ratio periodic structure in which the limitations of the process do not restrict the depth direction. Recent trends in the fabrication process for 2D high aspect ratio photonic crystals, which does not use a dry-etching process, are summarized in this section.

3.5.2 Glass capillary plate [1-3]

The capillary plate is the first material used to successfully produce a 2D photonic crystal in both the near infrared and the visible wavelength regions.[1] The core and the clad of the capillary plate are composed of glasses of various different refractive indices. They are arranged in a triangular lattice and then drawn out to reach the desired lattice constant. A hole array structure can then be obtained by selectively etching the core. Consequently, a glass capillary, the name of which comes from this fabrication process, of a hole array structure with an almost infinite aspect ratio can be fabricated. Although structures with a size of 0.2 μm as a hole periodicity have already been fabricated, there remain problems such as the generation of cracks caused by the strain that occurs during the core etching, when the hole period becomes smaller.

A review regarding 2D photonic crystals based on the capillary plate was included in *year 2000 report* by Masuda et al. Since then, an intensive study focusing on its application has been carried out. Inoue et al. stimulated a laser oscillation in a glass capillary with fine holes of a 0.67 μm period, which were filled with a dimethylsulfoxide (DMSO) solution coloring material, by pulse excitation with a YAG laser.[3] The reported low threshold input power for lasing of nearly 1×10^5 W/cm^2 is believed to arise via a high efficiency oscillation caused by an increasing interaction between the lightwave and the coloring material, due to a lowering of group velocity in the vicinity of the band-edge.

3.5.3 2D photonic crystal created by anodic etching of a semiconductor [4-7]

There is a method to fabricate a 2D photonic crystal by utilizing selective etching phenomena, which depends on the semiconductor crystal direction. These 2D photonic crystals can be fabricated from a high aspect ratio hole array. The anodic etching of single crystal semiconductors, formed from Si, InP, GaAs or a similar material, in an appropriate solution results in a selective etching process, where the selectivity depends on the nature of the crystal directions. Control of the hole generation position by means of lithography enables us to obtain a periodic hole array structure with a high aspect ratio.

After opening holes in the SiO$_2$ layer on a Si substrate by lithography, high aspect ratio holes, which are vertical to the (100) plane, are formed by anodic etching in HF electrolyte.[4,6] A high aspect ratio structure, which is difficult to obtain using the ordinary dry etching technique, can be obtained by the use of this anisotropic etching technique with a high selectivity ratio for a single Si crystal. Another study of the high aspect ratio hole array structure by anodic etching was reported with regard to compound semiconductors like InP and GaAs.[7]

3.5.4 Anodic of porous alumina [8-13]

After the anodic oxidation of aluminum in an acid electrolyte solution, a porous oxide film, which is composed of fine holes in a direction vertical to the film surface, is formed on the surface, as shown in Fig. 3.5.1. This covering film, known as anodic porous alumina, is the result of the formation of a regular structure, in which an amorphous state of alumina grows via self-organization, independent of the crystal direction.

Anodic porous alumina is considered to be one of the most promising materials for 2D crystals, because of the easy formation of a high aspect ratio hole array structure.[8] Hence, a combined technique, using an aluminum texturing process and the self-organization channel formation, has been developed as a fabrication method for anodic porous alumina, which

Figure 3.5.1 Porous alumina with an ideally ordered hole arrangement.

Figure 3.5.2 Fabrication process for porous alumina with an ideally ordered hole arrangement.

possesses an ideal fine hole arrangement, and is suited to 2D photonic crystals, as shown in Fig. 3.5.2.[10] Before the anodic oxidation, the aluminum surface is textured with a process that uses a mold with regularly arranged convexes. A regular arrangement of shallow concave depressions is formed on the aluminum surface as a result of this texturing process. These concave depressions help to induce hole generation during anodic oxidation and generate formation of a regular fine hole arrangement. A high yield can be obtained because the mold is made of a durable SiC wafer on which the convex structures are arranged regularly by an EB process. Consequently, it can be used repeatedly.

A measured example of the transmission spectrum of a 2D photonic crystal based on anodic porous alumina is shown in Fig. 3.5.3.[8] The stopband of transmission, which is common to both characteristic directions in the triangular lattice, Γ-X and Γ-J, is observed for s-polarized light and the formation of a 2D PBG in the visible wavelength range is verified. These results coincide well with the calculated results obtained by the plane wave expansion method.

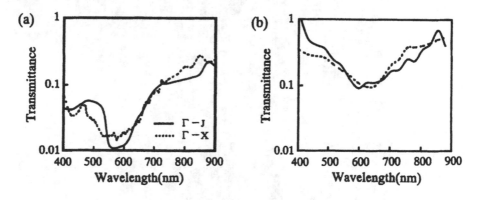

Figure 3.5.3 Transmission spectrum of a 2D photonic crystal of 200 nm period of holes. (a) s-polarization. (b) p-polarization.

It is desirable to form the regular fine hole arrangement with a large enough aspect ratio in the depth direction of the 2D photonic crystal. The anodic oxidation conditions are known to influence how well the ideal fine hole arrangement in the depth direction is maintained.[9] That is, when the anodic oxidation condition is not appropriate, the arrangement of fine holes falls into disorder, even if the texturing process induces a regular arrangement on the surface. In Fig. 3.5.4, an example of anodic porous alumina with a high aspect ratio formed under the appropriate anodization conditions after texturing is shown.[11] In the system depicted in Fig. 3.5.4, it

Figure 3.5.4 High aspect ratio for porous alumina with an ideally ordered hole arrangement.

Figure 3.5.5 Porous alumina formed by a self organization process.

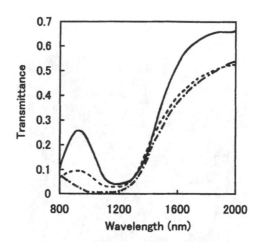

Figure 3.5.6 Transmission spectra of anodic porous alumina by a self-organization process.

is not necessary to use a microscopic spectrometer for evaluation of spectroscopic characteristics because the fine hole structure has an aspect ratio of over 200.

Mikulskas et al. reported that anodic porous alumina, which had been fabricated by using the texturing process before anodic oxidation, was utilized for controlling the position at which the fine holes were generated and hence to create a 2D photonic crystal. They obtained anodic porous alumina with a regular hole arrangement by pressing a grating, which was composed of grooves with a 830 nm period, onto the aluminum surface and by changing the angle at 60° intervals, so as to form a periodic texturing pattern on the alumina surface.[12] This resulted in a phenomenon where the cross points of the grooves formed by the change of angle became the hole-generating points.

The anodic porous alumina, in which fine holes are arranged regularly in a whole sample by the texturing process, is comparable with the hole array structure formed by the self organizing process, where a 2D PBG is also observed.[13] The regularly arranged structure of holes can be considered as a photonic polycrystal in this case because it possesses a domain structure with a size of several μm, whilst the anodic porous alumina, which possesses an ideal arrangement because of the texturing process, is considered to be a photonic single crystal. Figure 3.5.5 shows anodic porous alumina that has been formed by self-organization and Fig. 3.5.6 shows the transmission spectrum of this material. In the anodic porous alumina fabricated by the self-organizing process, the transmission light propagates through domains, which possess multiple crystal directions. The transmission spectra shown in Fig. 3.5.6 can be considered as the sum of the spectra for multiple crystal directions.

The theoretical treatment of 2D photonic crystals composed of domains and the effects of the domain boundary on light translation have not yet been investigated. However such a development would be attractive from the point of view of application, because of the ease of fabricating 2D photonic crystals with large areas.

3.5.5 Fabricating 2D photonic crystals by the replication processes

The 2D photonic crystal fabrication process, which uses anodic porous alumina as a template, is described. Various replication processes using anodic porous alumina as a template are being developed because this

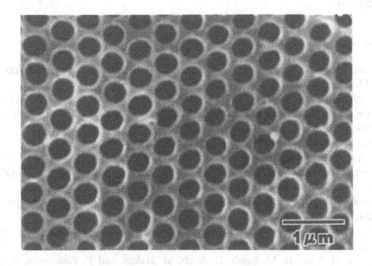

Figure 3.5.7 Hole array structure of LiNbO₃

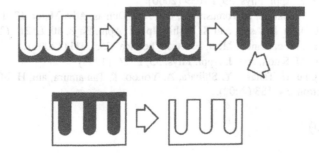

Figure 3.5.8 Two step replication processes for fabrication of hole array structure using an anodic porous alumina template.

substance has a relatively low refractive index (1.67) and photonic crystals have limited application at present, due to their lack of electric conductivity.

Figure 3.5.7 shows an example of a hole array structure in a $LiNbO_3$, which was fabricated using a template of anodic porous alumina. A 2D hole array possessing a geometric structure similar to the starting structure is obtained by means of a process that successively forms negative and positive structures, having started from anodic porous alumina, as shown in Fig. 3.5.8. The TiO_2 hole array structure can be fabricated by a similar process, and its PBG has already been evaluated.[14]

Anodic porous alumina can also be used as a mask for various replication processes by forming the through holes.[15] Nakao et al. used anodic porous alumina as an etching mask for an fast atomic beam and reported the fabrication of a waveguide formed from a fine process GaAs single crystal.[16]

References

1.   K. Inoue, M. Wada, K. Sakoda, A. Yamanaka, M. Hayashi, and W. Haus, Jpn. J. Appl. Phys. **33**, L1463 (1994).
2.   A. Rosenberg, R. J. Tonucci, H. B. Lin, and A. J. Campillo, Opt. Lett. **21**, 830 (1996).
3.   K. Inoue, M. Sasada, J. Kawamata, K. Sakoda, and J. W. Haus, Jpn. J. Appl. Phys. **38**, L157 (1999).
4.   V. Lehman, and H. Foll, J. Electrochem. Soc. **137**, 653 (1990).
5.   S. Rowson, A. Chelnokov, and J.M. Louritioz, J. Lightwave Technol. **17**, 1989 (1999).
6.   A. Birner, R. B. Wehrspohn, U. M. Gosele, and K. Busch, Adv. Mater. **13**, 377 (2001).
7.   T. Baba, and M. Koma, Jpn. J. Appl. Phys. **34**, 1405 (1995).
8.   H. Masuda, M. Ohya, H. Asoh, M. Nakao, M. Notomi, and T. Tamamura, Jpn. J. Appl. Phys. **38**, L1403 (1999).
9.   H. Asoh, K. Nishio, M. Nakao, T. Tamamura, and H. Masuda, J. Electrochem. Soc. **148**, B152 (2001).
10. H. Masuda, H. Yamada, M. Satoh, H. Asoh, M. Nakao, and T. Tamamura, Appl. Phys. Lett. **71**, 2770 (1997).
11. H. Masuda, M. Ohya, K. Nishio, H. Asoh, M. Nakao, M. Notomi, A. Yokoo, and T. Tamamura, Jpn. J. Appl. Phys. **39**, L1039 (2000).
12. I. Mikulskas, S. Juodkazis, R. Tomasiunas, and J. G. Dumas, Adv. Mater. **13**, 1574 (2001).
13. H. Masuda, M. Ohya, H. Asoh, and K. Nishio, Jpn. J. Appl. Phys. **40**, L1217 (2001).
14. H. Masuda et al., Proc. Annual Meet. Electro-Chemical Soc., 1J11 (2001).
15. H. Masuda, and M. Satoh, Jpn. J. Appl. Phys. **35**, L126. (1996)
16. M. Nakao, S. Oku, H. Tanaka, Y. Shibata, A. Yokoo, T. Tamamura, and H. Masuda, Opt. Quantum Electron. **34**, 183 (2002).

(by H. Masuda)

3.6 DEVELOPMENT AND APPLICATION OF 3D SEMI-CONDUCTOR COMPLETE PHOTONIC CRYSTALS

3.6.1 General Features

As has been already discussed in Section 2.2, the presence of a PBG in a photonic crystal causes light to behave in a manner different to that expected in conventional materials and therefore results in phenomena not known in conventional optics. For example, it was thought that spontaneous emission was an inevitable phenomenon, but spontaneous emission does not occur in the PBG, since the crystal has no allowed photonic states there. When point defects are artificially introduced into the crystal, the light becomes localized at the defects. Hence, the density of states increases around the defect level and a remarkable shortening of the emission lifetime follows and leads to the so-called thresholdless laser, which cannot be distinguished from either spontaneous emission phenomena or ordinary laser light.

The presence of the PBG makes it possible to isolate the atomic system from the outer field and enables pure interaction in the confined atomic system. By exploiting this principle, it is expected that the photonic crystal will be applied to the field of quantum computing drawing an increasing amount of attention to this field. It is also possible to guide and manipulate photons arbitrarily by introducing a number of artificial defects into photonic crystals, which leads to the production of photonic chip. As stated in Section 2.2 (also in 3.3 and 3.4), a 2D crystal slab can confine photons to a 3D space and important optical functions can be obtained. However, confinement in the vertical direction results from total internal reflection. This confinement is due to the difference of the refractive indices, therefore light leakage around the point defects is inevitable. The photonic chips with various optical elements, which can be produced by introduction of a number of defects into the crystals, require strong light confinement in the vertical direction. For this purpose, the realization of complete 3D photonic crystals is indispensable. A photonic chip[1] shown in Fig. 3.6.1 is one of the ultimate goals for photonic crystals. This figure indicates that ultralow threshold laser arrays with various oscillation wavelengths, waveguides with steep bends, optical modulators and wavelength splitters can all be integrated into an area as small as 100 μm^2 in a 3D photonic crystal by introducing the appropriate defects. The term *complete* is used to differentiate such structures from other *incomplete* crystals with leakage passes of photons for certain directions. Moreover, the complete crystal must satisfy the following requirements: it must be possible to introduce defects at arbitrary positions, to introduce high efficiency light-emitters, and to be electronically conductive.

Table 3.6.1 Effect of various fluctuations on the PBG of a stacked-stripe-type (or woodpile-type) 3D photonic crystal.

TYPE OF STRUCTURAL FLUCTUATION	BANDGAP WIDTH (%)	RATIO OF REDUCED BANDGAP TO THE PERFECT STRUCTURE
PERFECT STRUCTURE	15. 7%	1
MISALIGNMENT OF STRIPES OF X DIRECTION BY 20% OF ONE PERIOD	8. 8%	0. 56
MISALIGNMENT OF STRIPES OF X AND Y DIRECTIONS BY 20% OF ONE PERIOD	6. 6%	0. 42
DEVIATION OF INTERSECTING ANGLE BY 5°	15. 7%	1
VARIATION OF LAYER THICKNESS BY 20%	12. 8%	0. 82
VARIATION OF STRIPE WIDTH BY 20%	13. 3%	0. 85

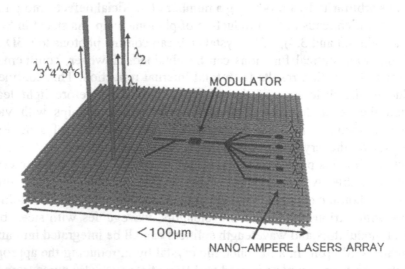

Figure 3.6.1 Example of photonic integrated circuit (or photonic chip) as one of the goals of photonic crystals. This figure explains that laser arrays of differing oscillation wavelengths with ultralow thresholds, a waveguide with a steep bend, an optical modulator and a wavelength splitter, can all be made in the very small area of 100 μm², by means of introduction of appropriate defects into 3D photonic crystals. The figure is a conceptual illustration. The idea of defect introduction leads to the expectation that it will be possible to produce numerous devices and circuits.

3.6.2 Method of Realization of complete 3D crystals

A variety of interesting 3D crystal fabrication methods have been proposed and tried, but most photonic crystal structures do not satisfy the above mentioned requirements (a complete PBG, the possibility of introducing defects and light-emitting material, and the ability to conduct current). For example, the 3D periodical structures (symmetrical fcc or simple hexagonal structure) formed by the opal method[2] and the autocloning method[3] lack a PBG. It is understood that a complete PBG can be formed in either a diamond structure or a fcc structure with asymmetric photonic atom.[4]

For the purpose of obtaining a complete photonic crystal that satisfies the above requirements, it was proposed to construct a 3D photonic crystal by stacking III-V semiconductor stripes using wafer fusion technique with a combination of laser-beam-assisted precise alignment.[1-8] The method is illustrated in Fig. 3.6.2. In this method, it is possible to (i) construct various types of structures and to introduce artificial defects at arbitrary positions, (ii) inject current into the crystal, and (iii) incorporate a high efficiency active region into the crystal structure itself. Thus, the method is considered to satisfy all of the requirements mentioned above.

The wafer fusion technique binds two wafers without the presence of any bonding agents in creation of crystal. This method is attractive because it enables us to not only integrate wafers of different lattice constant and different crystal structure,[9-11] but also to realize materials and devices with

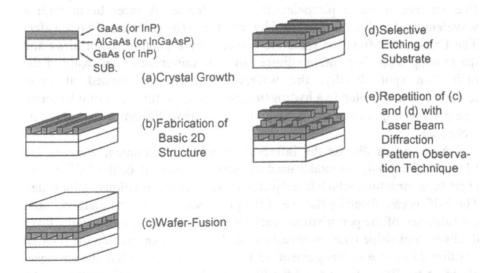

Figure 3.6.2 Method of realization of a complete 3D photonic crystal via the wafer fusion method.

entirely new structures, such as photonic crystals. The typical explanation of the wafer fusion technique and the fusion mechanism is as follows.[10] First, the surfaces of the wafers to be fused are cleaned and their natural oxide layer removed. They are then processed with a surface treatment to make them hydrophilic. When such hydrophilic wafers are laid upon each other, they bond via hydrogen bonding, even at room temperature. This bonding can be made stronger by heating, as dehydration will occur in the bond. Complete dehydration occurs at higher temperatures, when atoms at the interface interchange mutually and complete the fusion. The process for attaining this strong bonding is called wafer fusion.

The most important point in constructing photonic crystals is that each parallel stripe should be shifted by half a period, as shown in Fig. 3.6.2(e), to realize the diamond structure. A position alignment apparatus for this purpose was developed in order to observe the diffracted pattern of laser light, as is indicated in Fig. 3.6.3.[12] When a laser beam is injected into the III-V semiconductor stripes (which work as diffraction grating) in which the stripe patterns lie parallel to each other, the intensity of plus and minus first order diffraction spots changes according to its relative position between the individual stripes. The phenomenon is then utilized to align the stripes. Finally, the intensities of these first order diffraction spots become weakest when the relative position agrees with one half of the stripe period (in more detail, it may be influenced by both the distance and the multi-reflection between the diffraction gratings). To accomplish this, two wafers with stripes on top of them are held by wafer-holders, one of which is capable of fine movement via a piezoelectric effect device. A laser beam with a wavelength of 980 nm is diffracted by the stripes (or gratings) of both wafers. The first-order diffraction spots are detected by a photo-detector array and the moving holder's position adjusted so as to minimize the intensity of the diffraction spot. Finally, the wafers are hydrogen bonded at room temperature and heated in a hydrogen atmosphere for further strong bonding. The position alignment error is found to be less than 100 nm when using this technique.

Next, let us discuss the influence of position alignment error on the PBG. In Fig. 3.6.4, the model used to calculate the state of the PBG for an eight layer structure, which is subjected to various perturbations, is indicated. The PBG is calculated by the use of the plane wave expansion method,[13] and the influence of the perturbations such as relative position, cross angle, layer thickness and stripe width is investigated. Table 3.6.1 shows that the relative position difference (misalignment) in the x and y direction has the strongest effect on the PBG. The size of the PBG shrinks to about 42% of that in the perfect structure when misalignment reaches 20%. When a photonic crystal of the wavelength range for optical communications (wavelengths of around

1.55 μm) is fabricated under a condition of 100 nm misalignment, corresponding to a misalignment of about 14% and a PBG shrinkage of 60% compared to the perfect structure. In this case, the obtained net width of the PBG becomes 91 meV in terms of energy, which is sufficiently larger than room temperature energy. Therefore, misalignment error under 100 nm is within the permissible range.

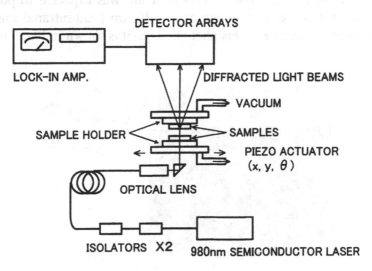

Figure 3.6.3 Outline of the alignment apparatus using a laser beam diffraction pattern observation technique.

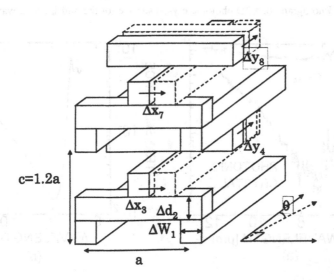

Figure 3.6.4 Model for calculating the fluctuation, which affects the PBG.

3.6.3 Development of 3D photonic crystals for the mid-infrared wavelengths

According to the method described in the previous subsection, a photonic crystal was fabricated by stacking four layers of semiconductor stripes of 4 μm periodicity, 1 μm width and 1.2 μm thickness.[1,6,7] The temperature used for the wafer fusion was 700°C. This structure was expected to possess a PBG within the wavelength range of 5 – 10 μm (mid-infrared range). A photograph of the surface of this photonic crystal is shown in Fig. 3.6.5. The

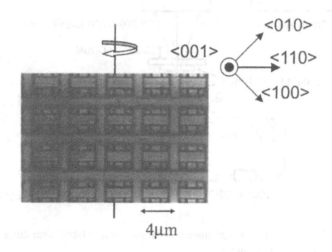

Figure 3.6.5 Photograph of a 3D photonic crystal surface for the mid-infrared wavelengths.

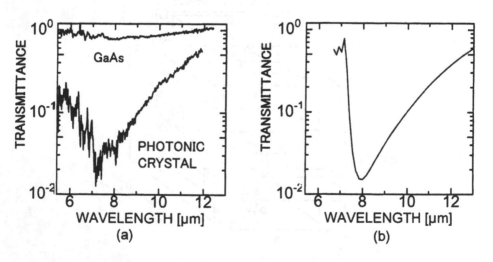

Figure 3.6.6 (a) Experimental results obtained for the transmission spectra of a crystal with four stacked layers. (b) Calculated result.

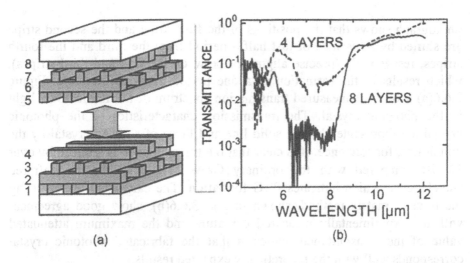

Figure 3.6.7 (a) Fabrication of an eight layer stacked crystal. (b) Transmission spectrum for the vertical incidence.

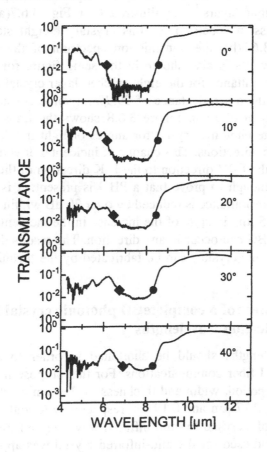

Figure 3.6.8 Transmission spectrum for various incidental angles.

photograph shows that the positions of the first stripe and the second stripe are shifted by a distance of just half a period from the third and the fourth stripes, respectively (precise alignment between the first and third stripes), which results in the formation of a one period diamond structure. Figure 3.6.6(a) shows the measured transmission spectrum of normal incident light in the photonic crystal. The transmission characteristics of the photonic crystal are represented by the solid line and those of a GaAs crystal by the dotted line, for reference. It is clear that the transmittance is reduced by over 16 dB compared with the ordinary GaAs crystal in the mid-infrared wavelengths, which indicates PBG formation. The calculated results using the transfer matrix method, shown in Fig. 3.6.6(b), show good agreement with the experimentally measured curvature and the maximum attenuated value of the transmittance, indicates that the fabricated photonic crystal corresponds well with the theoretically expected results.

Next, two photonic crystals, each consisting of four stacked layers, are stacked by the wafer fusion technique to produce a photonic crystal consisting of eight layers,[1,7] as illustrated in Fig. 3.6.7(a). The normal incident transmission spectrum for this crystal of eight stacked layers is shown in Fig. 3.6.7(b); the transmission spectrum of the individual four stacked layer crystals is also shown in the same figure for reference. The reduction in transmittance for the eight stacked layer crystal exceeds 30 dB, which corresponds to a reflectance of 99.9%, a value considered sufficient for the confinement of light. Figure 3.6.8 shows the transmission spectra obtained from the photonic crystal for angles of light incident between the <001> and <110> directions. This change of incidental direction corresponds to a range from the Γ-X' direction to the Γ-K direction of the band structure and is of itself enough to prove that a PBG is present. It is seen from this figure that the transmittance is reduced by over 20 dB within the wavelength range of $6.5 - 8.5$ μm, in spite of the increase in the incidental angle, which shows that the PBG can occur in any direction. The results demonstrate that satisfactory photonic crystals can be fabricated by the techniques described above.

3.6.4 Development of a complete 3D photonic crystal for optical communication wavelengths

The PBG wavelengths should be shortened in order to apply photonic crystals to optical fiber communications. For this purpose it is necessary to shrink the stripe period, width and thickness of the semiconductor stripes to around 0.70 μm, 0.17 μm and 0.19 μm respectively (about one fifth to one sixth of the size of a crystal for mid-infrared wavelengths). At first, the same fabrication method used for the mid-infrared crystal was applied to develop the crystal. The finished photonic crystal exhibits the PBG effect in the optical communication wavelengths, but its effect was one order of magnitude inferior to those of the crystals developed in the mid-infrared wavelengths. This is a critical problem for various applications.

To solve this problem, an investigation was implemented to discover which factors play an important role in the degradation of the photonic crystal characteristics. This investigation determined the two most important factors to be:

1. The collapse of stripe patterns at the time of wafer fusion, due to the mass transportation phenomena at the fused interface.

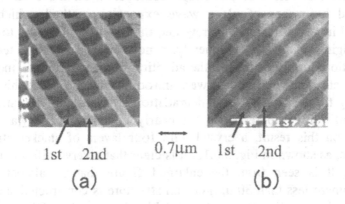

	1st 2nd	0.7μm	1st 2nd
	(a)		**(b)**

Figure 3.6.9 (a) Surface SEM photograph after stacking two layers, but before the process optimization for the near-infrared photonic crystal fabrication. (b) Surface photograph after the process optimization.

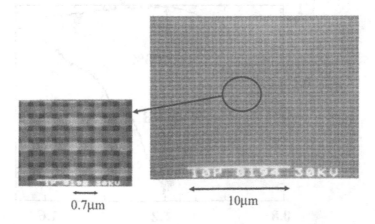

0.7μm   10μm

Figure 3.6.10 Surface SEM photograph of a complete four layer 3D photonic crystal for the near-infrared wavelengths.

2. The unexpected large change of shape that occurred during substrate removal, due to the over-etching process after the fusion of the two striped wafers.

Although the problems were not serious in large crystal structures for application in the mid-infrared wavelengths, it was found to be critical in a crystal for use in the near-infrared wavelengths. Figure 3.6.9(a) shows an SEM picture of the stripes after those processes. A significant degradation of stripe shape can be clearly seen in this figure. The stripe structure becomes round in shape and thinner than in other areas, particularly in the vicinity of the center. The effect of this shape degradation on the PBG effect was calculated by means of plane wave expansion method, which will be explained in next Chapter. This revealed that the PBG degrades to one tenth of its original width.[13] Consequently, a new method stressing temperature optimization during fusion and the adoption of an interference method into the substrate removing process were introduced to cope with this problem. Following these changes, shape degradation after stacking two stripe layers was significantly suppressed and a nearly complete shape was obtained. Based upon this result, a crystal with four layers of stacked-stripes was fabricated, as shown in Fig. 3.6.10. It is clear that a very uniform crystal was obtained. It is seen from the enlarged figure that an almost complete (misalignment less than 30 nm) crystal structure is constructed successfully in spite of the very fine stripe interval, which is as small as 0.7 μm.[8]

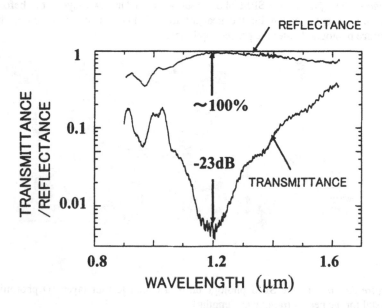

Figure 3.6.11 Transmission spectra of the normal incident for the crystal shown in Fig. 3.6.10.

The measured results of the transmission and reflection spectrum obtained by the vertical incidence into this structure are shown in Fig. 3.6.11. The figure shows a large transmission attenuation of –23 dB, indicating the existence of the PBG and >99% reflection is realized despite only using a

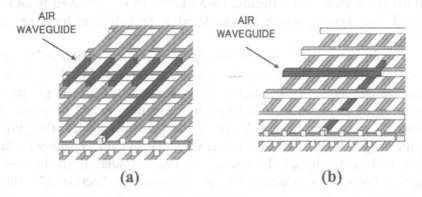

Figure 3.6.12 (a) Typical model of a waveguide with a right-angled bend using one layer as a defect layer. (b) Right-angled bend waveguide using two layers.

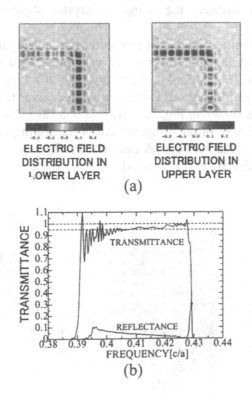

Figure 3.6.13 (a) Light guiding results calculated by the FDTD method for the structure shown in Fig. 3.6.12(b). (b) Transmission spectrum for the bend.

four layer stack. These results also show the optically perfect characteristics of the crystal. The measurement of the transmission spectrum from various incidental directions of light leads to the discovery of a complete PBG in the 1.3 – 1.55 μm range. Moreover, an almost ideal transmission attenuation of –40 dB (equivalent to a reflection ratio of 99.99%) is achieved in an eight stacked layer crystal, which is considered to be sufficient for a photonic crystal.[8]

The fundamental studies on 3D crystals have made the application of these structures to photonic chips feasible, as is shown in Fig. 3.6.1. In order to realize the chips, further steps such as element design, test production, evaluation of characteristics and circuit integration should be taken. The theoretical guidelines for the design of 2D photonic crystals that possess infinite heights are already set, but in reality, 2D photonic crystals with finite height will be fabricated. Therefore, the loss of radiation in the vertical direction becomes a critical issue, as discussed in Section 2.2. Hence, considerable developments are required to realize bend waveguides with a low loss for a wide wavelength range.[14] The accurate design and fabrication of a waveguide structure using a 3D crystal will allow us to realize such bend waveguides because the complete crystal structure is free from radiation loss.

Although numerous degrees of freedom exist in designing a waveguide that uses 3D crystal, the limitations with crystal fabrication mean that it is preferable to use as few stacked layers as possible in waveguides. As an example, one layer of the crystal is used as a defect layer to form the bend waveguide. This is shown in Fig. 3.6.12(a). In this case, the bend loss is relatively large because of the disagreement of translating modes existing at the front and rear of the 90° bend. A calculated reflection loss of about 50% is found by using the 3D FDTD method.

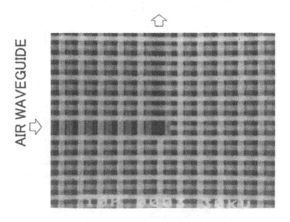

Figure 3.6.14 SEM photograph of the fabricated waveguide with a right-angled bend.

Through various trials and errors, the most effective structure for the bend waveguide in 3D crystal was found to be a structure using two photonic crystal layers in which one missing stripe is set with crossed configuration. In this case, the waveguide modes coincide with each other before and after the bend,[15] as is shown in Fig. 3.6.12(b). The optical transmission behavior calculated by the 3D FDTD method is shown in Fig. 3.6.13(a). Light propagates through the waveguide formed in the lower layer before the bend. When it reaches the bend corner, the light moves to the upper layer and then propagates through the waveguide formed in the upper layer, as shown in the figure. The effectiveness of this bend waveguide can be demonstrated by its transmission spectrum, which indicates a transmittance of over 95% for a wide wavelength range, as depicted in Fig. 3.6.13(b). These ideas can be applied to various optical circuits. For example, an ultrasmall T-type splitting guide can be made by simply extending the length of one side of the guide. An actual bend waveguide based on the design mentioned above[15] was successfully constructed as shown in Fig. 3.6.14. Attempts have already been made to introduce a light-emitter into a 3D photonic crystal. It was observed that photonic crystals will have a large effect on light emission phenomena.[16] These results are believed to be a first step towards the realization of an optical chip in the future.

References

1. S. Noda, N. Yamamoto, M. Imada, H. Kobayashi, and M. Okano, J. Lightwave Technology **17**, 1948 (1999).
2. V. Fornes, and A. Mifsud, Appl. Phys. Lett. **71**, 1148 (1997).
3. S. Kawakami, Electron. Lett. **33**, 1260 (1998).
4. K. M. Ho, C. T. Chan, and C. M. Soukoulis, Photonic Band Gaps and Localization (Ed. by C. M. Soukoulis), Plenum Press, 235 (1993).
5. S. Noda, N. Yamamoto, and A. Sasaki, Jpn. J. Appl. Phys. **35**, L909 (1996).
6. N. Yamamoto, S. Noda, and A. Chutinan, Jpn. J. Appl. Phys. **37**, L1052 (1998).
7. S. Noda, N. Yamamoto, H. Kobayashi, M. Okano, and K. Tomoda, Appl. Phys. Lett. **75**, 905 (1999).
8. S. Noda, K. Tomoda, N. Yamamoto, and A. Chutinan, Science **289**, 604 (2000).
9. J. B. Lasky, Appl. Phys. Lett. **48**, 78 (1986).
10. M. Shimbo, K. Furukawa, and K. Tanzawa, J. Appl. Phys. **60**, 2987 (1986).
11. H. Wada, Y. Ogawa, and T. Kamijoh, Appl. Phys. Lett. **62**, 738 (1993).
12. N. Yamamoto, and S. Noda, Jpn. J. Appl. Phys. **37**, 3334 (1998).
13. A. Chutinan, and S. Noda, J. Opt. Soc. Am. B **16**, 240 (1999).
14. A. Chutinan, and S. Noda, Phys. Rev. B **62**, 4488 (2000).
15. A. Chutinan, and S. Noda, Appl. Phys. Lett. **75**, 3739 (1999).
16. M. Imada, S. Noda, A. Chutinan, T. Tokuda, H. Kobayashi, and G. Sasaki, Appl. Phys. Lett. **75**, 316 (1999).

(by S. Noda)

3.7 AUTOCLONED PHOTONIC CRYSTALS AND THEIR APPLICATION FOR VARIOUS DEVICES

3.7.1 The autocloning method

The photonic crystal is an innovative artificial optical material, which has a periodically arranged structure, produced from materials of different refractive index. It has unique characteristics such as strong dispersion and anisotropy in the pass band and a cut-off frequency imposed by the PBG.[1] It is attracting much attention in fields of optics such as fiber communication as a new technique for developing next-generation devices, and practical applications are expected to appear soon.

The key to producing photonic crystals for practical use is a simple and highly reliable fabrication technique yielding a multi-dimensional periodic structure of a submicron periodicity with high quality. A simple method named as autocloning technique, which is based on radio frequency (rf) bias sputtering, has been developed and various kinds of photonic crystals and functional devices are being developed.[2] In this section, the fabrication technique is outlined, and polarization splitters for perpendicularly incident light and a functional optical circuit for propagation in the in-plane direction are introduced as device applications. Commercial manufacturing of the polarization splitters will soon commence and will be the first application of autocloned photonic crystals. Functional optical circuits are expected to be used for next-generation planar lightwave circuits (PLCs) because low loss propagation and high coupling efficiency to fibers have been verified, and new functions, which cannot be realized in conventional silica-based PLCs, have been demonstrated.

The concept of the process is shown in Fig. 3.7.1. After the formation of the periodic corrugation pattern on a substrate by EB lithography, the layers are stacked on the substrate by a combined sputtering deposition and sputter etching process. Under the appropriate conditions, the deposition of the multilayer progresses with the preservation of the surface corrugation pattern. In this way, a multi-dimensional periodic structure (a 3D or 2D periodic structure with a periodicity along the stacking direction) can be formed sequentially. Figure 3.7.2 shows a cross-sectional SEM photograph and a surface atomic force microscope (AFM) image for the autocloned photonic crystals composed of Si and SiO_2. During the stacking process, the surface shape automatically evolves into a triangular wavy shape, despite the rectangular pattern formed on the substrate. The surface shape is stable against disturbances because it is determined by the process conditions[3]. Therefore, this method is reproducible, reliable and suitable for industrial production. Moreover, this method has a higher degree of freedom regarding

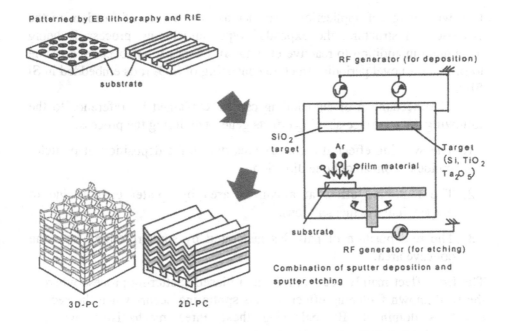

Figure 3.7.1 Concept of the autocloning process.

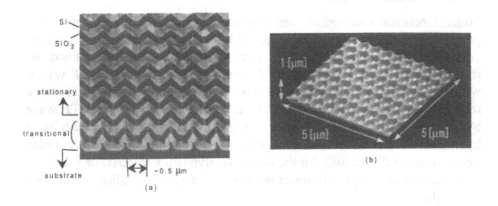

Figure 3.7.2 (a) SEM image of the cross-section of a 2D autocloned photonic crystal. (b) AFM image of a 3D autocloned photonic crystal.

the selection of materials and the choice of in-plane patterns. For example, not only Si but also visible-light-transparent materials such as Ta_2O_5 or TiO_2 and functional materials can be selected. It is also possible to select a variety of patterns, such as straight and curved grooves and holes arranged in a triangular or square lattice. Such flexibility enables this method to be applied to a wide range of applications. To increase the degree of freedom in the cross-section structure, the expanded type autocloning process is being developed in addition to reactive etching, and hence it has become possible to produce a novel periodic structure consisting of SiO_2 rods embedded in Si film.[4]

The mechanism of autocloning can be explained by reference to the following three surface-shaping effects generated during the process.[5]

1. The swelling effect at a convex area due to the deposition of particles incident from an oblique direction.

2. The removing effect at a convex area by sputter etching due to perpendicularly incident ions.

3. The re-deposition of particles generated by sputter etching onto the concave area.

The first effect mainly occurs in the non-biased sputtering process whereas the well-known flattening effect by bias sputtering occurs when the second effect is dominant. By balancing these three mechanisms, we can consistently duplicate corrugation patterns.

3.7.2 Polarization splitter

Some functional components are examined as applications of autocloned photonic crystals for normal incidence of light. Here, the polarization splitter composed of 2D photonic crystal, as shown in Fig. 3.7.3, is explained. As the stopband and the pass band are different for the TM and the TE waves, which are polarized perpendicular and parallel to the groove direction respectively, the transmitted TM wave can be separated from the TE wave by reflection.[6] To obtain a wide operation range, a combination of a hydrogenated amorphous silicon (a-Si:H) for the high refractive index material ($n = 3.5$) and SiO_2 for the low refractive index material ($n = 1.45$) is suitable, due to a large difference between the refractive indices of the two materials.

The results of the measurement of the transmission spectrum of the fabricated structure are shown in Fig. 3.7.4. It operates as a polarizer for the wavelength range of $1.4 - 1.6$ µm. In detail, the insertion loss of less than 0.15 dB for the TM wave and the extinction ratio of more than 50 dB for the TE wave were measured in the wavelength range from 1.52 µm to 1.58 µm. Both of these values are sufficient for practical use.[7] An anti-reflection coating is formed by selecting appropriate materials and thicknesses for the first and last layer.

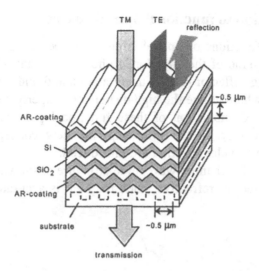

Figure 3.7.3 Structure of the photonic crystal polarization splitter.

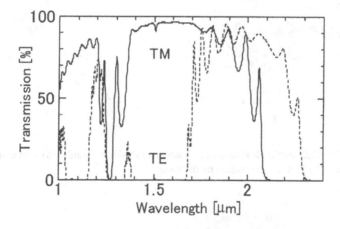

Figure 3.7.4 Transmission spectrum of a photonic crystal polarization splitter.

This polarizer possesses many advantages over conventional designs including its high performance, thin, large aperture, light weight, low material costs, and a surface that does not require polishing. The functional material can be used as the substrate, for example, it has proved possible to develop a polarizer stacked directly on a garnet crystal.[8] The process for achieving high uniformity over a large area is being developed so that it may be mass-produced. A monolithically integrated isolator and a polarized beam combiner module for a Raman amplifier are promising applications.

3.7.3 Application to functional optical circuits

More complex functions of optical circuits can be realized by propagating light along the plane of the autocloned photonic crystal. Functions such as distributed Bragg reflector (DBR) using a stopband and a wavelength filter or a delay line, which utilize the high dispersion property near the band-edge, are possibilities. Combination of these functional parts with a waveguide structure is necessary. In the following, the new concepts of a channel waveguide and a wavelength filter are introduced.

An example of a channel waveguide structure is shown in Figs. 3.7.5(a) and (b).[9] The effective refractive index for light is accurately controlled by

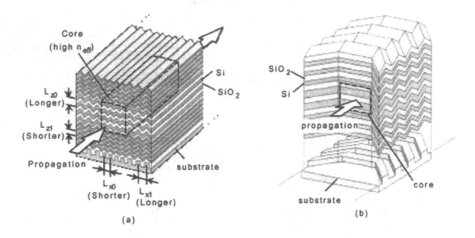

Figure 3.7.5 Schematic illustration of photonic crystal waveguide. (a) Lattice-constant modulation. (b) Lattice-orientation modulation.

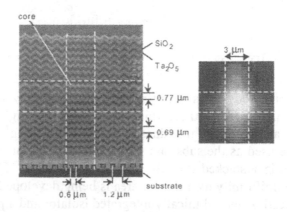

Figure 3.7.6 Near-field pattern of a lattice-modulated waveguide at $\lambda = 1.55\ \mu m$.

changing the period of the stacking layers and the period and orientation of the in-plane pattern (lattice modulation). Thus, a channel waveguide, as shown in Fig. 3.7.5, can be realized by forming a core with a highly effective refractive index. Modulation of the lattice constant is shown in (a), whilst modulation of the lattice orientation is shown in (b). The cross-section and the near-field pattern of the lattice constant modulated waveguide fabricated from a Ta_2O_5/SiO_2 system are shown in Fig. 3.7.6.[10] The boundaries between the areas with a different in-plane pitch are found to grow in the direction perpendicular to the substrate. The spot diameter is about 4.5 μm. The fiber-to-fiber transmission measurement of a photonic crystal waveguide has been accomplished by butt-jointing. Butt-jointing involves the direct connection of the waveguide to a high Δ single mode fiber, which is spliced by the tapered expanded core (TEC) technique to a standard SMF. The total loss of the fiber-to-fiber configuration in a waveguide of 3 mm length is shown in Fig. 3.7.7. A loss of 4.6 dB at minimum is obtained, which includes the coupling loss due to the Fresnel reflection and the mode mismatch. As the coupling losses can be estimated to be about 0.7 dB in total, the transmission loss is calculated to be about 1.3 dB/mm. The good connectability with fiber and the low transmission loss are thus experimentally verified.

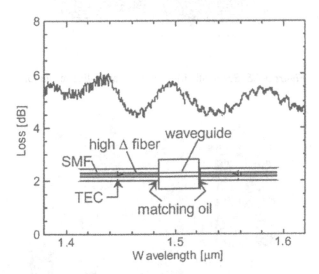

Figure 3.7.7 Fiber-to-fiber loss spectrum of a 3 mm-long photonic crystal waveguide.

Various functions can be realized in an optical circuit based on a lattice modulation waveguide. For example, the optical resonator shown in Fig. 3.7.8 was fabricated and evaluated[10]. The resonator is composed of mirrors and a cavity, which operate as PBGs for the incident wavelength and a pass band, respectively. Both ends of the resonator are connected to the channel waveguide, as shown in Fig. 3.7.5(a) for example. In the resonator, the light is confined within the core area, which has a high effective refractive index and is formed by the large period of stacking layers, similar to lattice modulation waveguide. Figure 3.7.9 shows the pattern on the substrate for a resonator with waveguide and a fabricated resonator after stacking Ta_2O_5/SiO_2 multilayer on the substrate. The fabrication is simple since the 2D or 3D periodic structure, which is a heterostructured photonic crystal with differing periods and directionalities, is formed by single stacking of a multilayer on the patterned substrate. There is no need for any alignment. The transmission spectrum of an optical resonator with a waveguide of 1.5

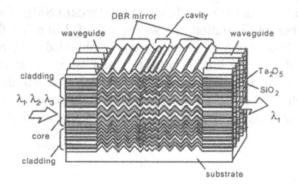

Figure 3.7.8 Schematic illustration of an in-line resonator.

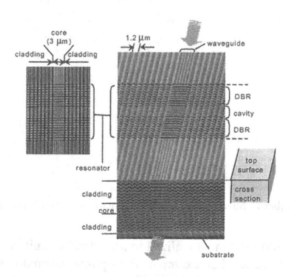

Figure 3.7.9 SEM images of (a) a patterned substrate and (b) an in-line resonator (perspective).

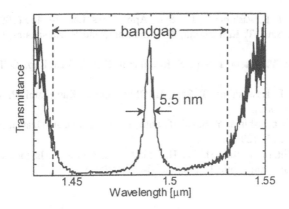

Figure 3.7.10 Transmission spectrum of an in-line resonator (Q = 270).

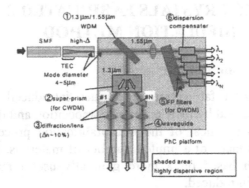

Figure 3.7.11 Concept of a photonic circuit based on autocloned photonic crystal.

mm length was measured by the fiber-to-fiber system, as shown in Fig. 3.7.10. A resonating peak width of Q = 270 was obtained at a wavelength of 1498 nm. The center frequency and the Q-value are adjustable in the initial patterning process. As the conventional PLCs (planar lightwave circuit) cannot form such an in-line type resonator structure, this example shows the high functionality of autocloned photonic crystal circuits. The next target may be to realize photonic crystal optical integrated circuits with various functions, as shown in Fig. 3.7.11.

References

1. S. Kawakami and T. Sato, Proc. European Conf. Optical Commun., Tu B3.1 (1999).
2. S. Kawakami, Electron. Lett. **33**, 1260 (1997).
3. T. Kawashima, K. Miura, T. Sato, and S. Kawakami, Appl. Phys. Lett. **77**, 2613 (2000).
4. T. Kawashima, T. Sato, Y. Ohtera, and S. Kawakami, IEEE J. Quantum Electron. **38**, 899 (2002).

5. S. Kawakami, T. Kawashima, and T. Sato, Appl. Phys. Lett. **74**, 463 (1999).
6. Y. Ohtera, T. Sato, T. Kawashima, T. Tamamura, and S. Kawakami, Electron. Lett. **35**, 1271 (1999).
7. T. Kawashima, Y. Ohtera, T. Sato, S. Kawakami, IEICE Tech. Rep., OPE99-109 (1999, in Japanese).
8. W. Ishikawa, T. Kawashima, T. Sato, H. Ohba, and S. Kawakami, Proc. IEICE General Conf., C-4-6 (2002).
9. Y. Ohtera, T. Kawashima, Y. Sakai, T. Sato, I. Yokohama, A. Ozawa, and S. Kawakami, Opt. Lett. **27**, 2158 (2002).
10. T. Sato, Y. Ohtera, T. Kawashima, H. Ohkubo, K. Miura, N. Ishino, and S. Kawakami, Proc. European Conf. Optical Commun., 04.4.2 (2002).

(by T. Sato)

3.8 PHOTONIC CRYSTALS FABRICATED BY MICROMANIPULATION METHOD

3.8.1 Micromanipulation method

The photonic crystal fabrication techniques introduced in the previous sections were classified by their material composition and dimension (2D or 3D). This is because most of the microfabrication processes cover only specific structures made of a limited number of materials. In this section, a different approach based on a more generally useful micromanipulation technique will be introduced.

Various techniques have been established in recent years for the fabrication of optical wavelength 3D photonic crystals. The method using a

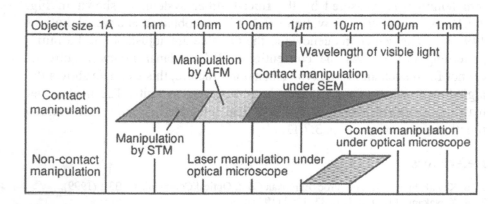

Figure 3.8.1 Various micromanipulation techniques.

precise semiconductor process is very promising.[1,2] However, successful demonstrations using this method have been limited, since it demands several sequences of highly controlled processes. Another technique, utilizing the self-organization of colloidal particles[3] enables us to realize a large scale crystal at relatively low cost, but neither a completely defect-free single crystal nor a crystal with designed defects have been produced. No other methods are yet sufficient for fabricating a complete structure with the necessary optical characteristics. Conversely, the micromanipulation method is able to realize a completely controlled artificial 3D structure at the dimensions of optical wavelengths by means of the mechanical assembly of fine parts made of colloidal particles or by the semiconductor technique. Even though this technique may only be applied to small-scale crystals, the method is well suited for the systematic investigation of characteristics and for the exploration of new phenomena, as it can produce crystals flexibly without dependence on either the material or the structure. Furthermore, by taking into consideration the fact that recent high density circuits for portable computers or cellular phones are mass-produced by an ultrahigh-speed automatic assembly technique, based on image recognition under an optical microscope, it will soon be possible to mass-produce large photonic crystals by automatic manipulation.

The three different micromanipulation techniques are as follows:[4] The first is a method to combine an optical microscope or a scanning electron microscope (SEM) and the micromanipulator, the second is a way to use a scanning probe microscope for both the observation and the operation, and the third is a method to use the radiation pressure of a laser for object control. Figure 3.8.1 shows the coverage of objective size that each method can treat. Of these methods, manipulation under the observation of an optical microscope or an SEM is most suitable for the assembly of a 3D microstructure.

Biologists frequently perform manipulation under the observation of an optical microscope (stereoscopic or inverted microscopes are usually used). Micromanipulators for optical microscopes and various kinds of peripheral apparatus are commercially available, and there is much accumulated knowledge of this area. It becomes indispensable to use an SEM for observation when the size of the object decreases to several μm or less. An SEM is most suitable for manipulation of an object size of optical wavelength as the SEM has an image resolution of several nm and a deep focal depth. There are, however, many restrictions because the micromanipulators for SEMs are extremely limited and all operations must be executed in a narrow vacuum chamber. The reason for the unpopularity of the SEM manipulators is not a technical difficulty, but rather the reluctance of manufacturers to commercialize these devices due to a lack of profitability. Development of a home-built micromanipulator for an SEM is no longer difficult because a positioning mechanism with a high accuracy and a compact size for use in a vacuum is readily available. Thus, such a manipulator will necessarily become popular as a general research tool in the near future. In this section, the arrangement of microspheres and the stacking of semiconductor microplates under SEM observation are introduced.

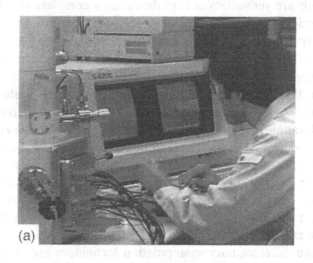

(a)

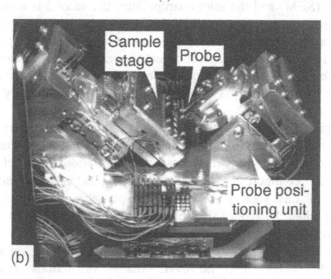

(b)

Figure 3.8.2 Micromanipulation systems. (a) Micromanipulation scene. (b) Micromanipulator.

3.8.2 Micromanipulation system under SEM observation

A system with a field emission type SEM (Hitachi S4200) which is coupled with a piezoelectric type micromanipulator (NanoRobot II), is shown in Fig. 3.8.2. This manipulator is composed of a sample stage and a probe positioning unit. It can determine the probe tip position at an arbitrary point within a 15 mm cubic area on the sample stage with an accuracy of 10 nm by means of a unique coarse and fine positioning mechanism.[5] An operator maneuvers this system by joysticks whilst observing a real-time image with a magnification from 5 000 to several tens of thousands.

The probe tip makes contact with and carries the fine object. The probe is usually made of a thermally drawn glass needle, which is coated with Au. The object adheres to the probe tip by electrostatic forces or Van der Waals forces,[6,7] and this adhesion force depends strongly on many conditions, such as the material combination, the curvature radius of the probe tip, the electron beam condition and especially the acceleration voltage. Under the appropriate conditions, the object can be reversibly picked up or deposited just by properly choosing the trajectory of the probe. As necessary, the adhesion force to the substrate is increased by applying a polymer film coating on the substrate, or the adhesion force to the probe can be increased by applying a voltage of several tens of volts to the probe. It has been confirmed that micro-objects with various shapes, such as spheres, whiskers, and plates made of various materials, such as metals, polymers, ceramics, and semiconductors, can be manipulated in this way. A metallic sphere with a diameter of 70 nm is the smallest object manipulated to date.[8]

3.8.3 Fabrication of 2D and 3D photonic crystals by microsphere arrangement

Before micromanipulation, a suspension of microspheres in an appropriate solvent is dropped onto the substrate and dried. This distributes the microspheres on the substrate such that they do not interact with each other. Thus it is not necessary to apply an electroconductive coating to the spheres, even when they are made of insulating materials. This is because when an electron beam, which is accelerated to several kV, is injected into micro-objects with a size below several μm, most electrons penetrate the object and little electrification occurs. Therefore, there is no hindrance to SEM observation. Polystyrene and silica spheres with a standard deviation in diameter of about 1% are commercially available as mono-dispersed microspheres. However, as this deviation is sufficiently large to prevent fabrication of a defect-free lattice, the most appropriate spheres are chosen by measuring each sphere's size and shape on the SEM image. (The

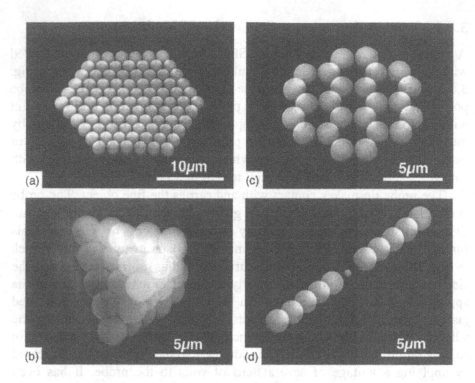

Figure 3.8.3 SEM micrographics of typical microstructures by the mechanical arrangement method;[6,9] (a) 2D close-packed structure. (b) 3D fcc structure. (c) Honeycomb structure. (d) Photonic molecule.

permissible level of deviation depends on the experimental purpose. For example, a typical limit may be within ±0.5%). Moreover, in order to arrange the spheres with high accuracy, a computer-aided-design (CAD) image is superimposed on the SEM monitor, and the spheres are positioned to agree with the CAD image.

A requirement for SEM observation is that the substrate must be electroconductive or electron-beam-transmissive. Usually, an electroconductive substrate, which is comprised of an evaporated indium tin oxide film on glass or an electron-beam-transmissive SiN membrane, with a thickness of several 100 nm is used. However, careful selection of acceleration voltages enables microsphere manipulation even on an insulating substrate. Figure 3.8.3 shows examples of microsphere structures fabricated in this way. The actual scale of the product is the cluster size, rather than the crystal size, as is seen from these photographs. It is demonstrated that as a result of (b) a 3D lattice can be formed, and as a result of (c) that the intended defect can be inserted as designed, and finally, as a result of (d) that a complex structure, which is composed of different kinds of particles (polystyrene with a diameter of 2 μm and silica with a diameter of 0.6 μm) can be fabricated. It required 8 hours to arrange the 91 particles of the structure (a). It took just a few seconds to pick up, move, and then deposit the spheres on the substrate. Almost all of the time was spent searching for usable spheres, in aligning the positions, and in the confirmation of arrangement accuracy.

As one example of the research in this field, the observational results for the process in which the photonic band grows as the cluster becomes larger in a close-packed monolayer arrangement of polyvinyltoluene spheres (diameter of 2 μm and refractive index of 1.6) are introduced.[9] The transition of transmission spectra for normal incidence, when the number (N)

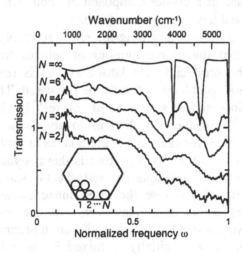

Figure 3.8.4 Relationship between cluster size and transmission spectra.[9]

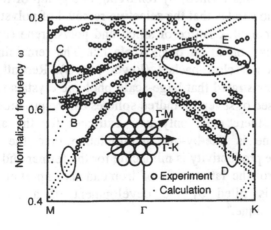

Figure 3.8.5 Photonic band diagram.[9]

of spheres at the edge of a hexagonal arrangement is gradually increased is indicated in Fig. 3.8.4. The deep and steep growth of the two valleys is seen in the figure. $N = .$ corresponds to the results of the calculation for infinite arrangement by the vector spherical wave expansion method and the valleys correspond to photonic bands. Figure 3.8.4 shows the gradual build-up of the photonic bands. If the incidence angle of light is changed, the valley separates or shifts. In Fig. 3.8.5, the valley positions according to the incidence angle are plotted on the dispersion diagram for a cluster where $N = 6$ (Fig. 3.8.3(a)). A relatively good agreement with the result of the calculation is seen in the experimentally obtained band diagram. The finite sphere number causes this agreement to deteriorate at bands B, C, and E. However, in the case of a cluster composed of about 100 spheres, a rough form of photonic band has already been formed.

There are two problems regarding microsphere arrangements produced by manipulation: that the usable number of spheres for arrangement is limited and that the producible 3D lattice system is restricted. A recent attempt to solve these problems will now be discussed. The complete PBG does not open in close-packed structures of spheres[10] but it opens widely in diamond structures.[11] However, it is difficult to fabricate a diamond structure of microspheres by the self-organization method, as the volume fraction of a diamond structure is small, so the structure is thermodynamically unstable. Even if the manipulation technique is used, the fabrication of a diamond structure is still difficult because the interconnection technique between spheres is not yet established. However, we can use the fact that superposition of two diamond structures with an appropriate shift makes a body-centered-cubic lattice. Initially a mixed body-centered-cubic-lattice (mbcc) was produced from particles of two different types. Subsequently, a diamond structure was formed, by removing one group of these particles.[11] Figure 3.8.6 demonstrates that the selective removal of polystyrene (denoted by X) from the complex lattice of silica and polystyrene (upper panels) is possible via utilization of an oxygen plasma. The remaining structure of silica spheres is not close-packed (lower panels). Recently, it has been successfully demonstrated that large-scale photonic crystals with a diamond structure composed of several hundred spheres can be produced.[12]

At present, there exist only two extremes for the arrangement of microspheres; one is one-by-one manipulation, and the other is self-organization. The productivity is miniscule for the former and it is extremely difficult to control the latter. It is of tremendous importance that the gap between them is filled by the development of a new microsphere arrangement technique.[4]

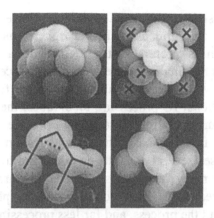

Figure 3.8.6 Selective etching scene of a polymer particle by oxygen plasma.[12]

3.8.4 Fabrication of 3D photonic crystals by microplate stacking

Since the concept of a PBG was proposed in 1987, researchers have sought to realize optical integrated circuits, in which low threshold lasers,[13] sharp-bend waveguides and multiplexers are accumulated on one chip.[14] 3D wiring and arrangement of elements are adapted in electronic devices in order to realize high integration, thus we can easily imagine that optical integrated circuit would also require a 3D arrangement of waveguides and active elements. In order to fulfill such a requirement, 3D photonic crystals, which can control light in all directions, seem to be indispensable. However, the technical requirements for 3D photonic crystals, such as structural accuracy in the order of tens of nanometers and high flexibilities in design and materials, exceed the capability of the present ultrafine processing technology. Various fabrication techniques, such as stereoscopic etching,[15] colloidal precipitation,[15-19] photopolymerization,[20-22] and stacked-stripe,[1,2,23,24] have already been proposed and stacked-stripe techniques that make excessive use of semiconductor ultrafine processing technologies[1,2,23,24] are considered to be the most promising methods at present. The production of optical wavelength crystals with a few periods and the introduction of a simple defect were achieved by these techniques. However, the complicated procedure in addition to damage inflicted upon the fine structure caused by frequent dry etching, polishing, and heating rendered further multilayering and the introduction of various materials and defects difficult. Furthermore, the integration of multiple components has yet to be achieved.

As we have introduced in Subsections 3.8.1 and 3.8.2, the techniques for microsphere arrangement have been well established enough to produce large defect-free crystals and designated defects. However, we can not deny that the way to use zero-dimensional microspheres as unit components is counterproductive. It is preferable to begin with a higher dimensional component to obtain the final structure with fewer processes. In this section, 2D photonic plates, which serve as unit components for 3D crystals, were prepared by conventional IC processing techniques, then, these plates were assembled by micromanipulation. To achieve the correct lamination of each layer, we adapted a fiducial shape and matching component as a stopper. This method possesses several advantages compared to the aforementioned techniques. Firstly, many 2D plates can be prepared by a single IC processing procedure. Thus, good accuracy of the photonic structure is maintained throughout the process, and far less processing is required than in stereoscopic etching and layer-by-layer methods. Secondly, the accurate alignment of lattices is automatically achieved regardless of photonic patterns or materials. Thus, any photonic structure can be assembled, regardless of how complicated the lattice structure and materials are. Thirdly, micromanipulation is very good when handling materials of sizes between 100 nm and several hundred microns, thus, small optical devices can be introduced into the structure at arbitrary positions. In short, the technology that have been presented in this section has important features that are lacking in existing technologies. Here, fabrication technique of the quasi-face-centered (stacked-stripe or woodpile) structure is demonstrated, which is a very well known complete PBG structure. The crystals obtained are evaluated from the standpoint of their optical properties.

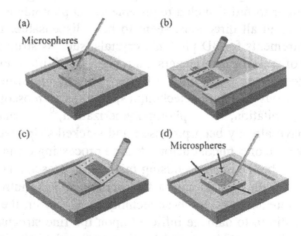

Figure 3.8.7 Schematics of plate assemblage. (a) Insertion of microspheres into fiducial holes of a base. (b) Separation of a plate by breaking bridges with a probe. (c) Superposition of a plate onto the substrate with microspheres. (d) The plate is fixed on the base by inserting another microsphere into a residual hole, and two more spheres are inserted for the next stacking.

Figure 3.8.7 shows a schematic diagram of the fabrication process. The unit plates for 3D photonic crystals were supplied as airbridge plates sustained in the air by narrow bridges. As the material for the photonic plate, InP was chosen because semiconductors can be directly applied for practical use. Details of the fabrication process are indicated in Ref. 25. Here, we adapted fiducial holes, which were prepared in the frame of photonic plates, and polystyrene microspheres as stoppers. First, two polystyrene microspheres were inserted into the fiducial holes of the base using a fiber probe (Fig. 3.8.7(a)). Then, the probe was exchanged with a thicker one for plate assembly. The bridges of the plates were broken by poking the notch of each bridge with a probe tip (Fig. 3.8.7(b)). By tuning the accelerating voltage of the SEM, the affinity of the material to the substrate or the probe can be controlled. When the plate was superposed on the base, the microspheres guided the plate to the proper position (Fig. 3.8.7(c)). Then other microspheres were inserted into fiducial holes for the next plate assembly (Fig. 3.8.7(d)), and this procedure was repeated to obtain piled structures. Figures 3.8.8(a) and (b) show bird's eye views of completed four (one period) and eight (two periods) layered crystals. The size of the crystals are $25 \times 25 \times 2$ μm^3 and $25 \times 25 \times 4$ μm^3, for four and eight layered crystals, respectively. No interstice between layers was observed in the SEM images. From the top view of the eight-layered crystal (Fig. 3.8.8 (c)), we can confirm that the accuracy of periodicity was retained, regardless of the increase of the number of layers, and that the periodical error was kept within 50 nm.

Reflection and transmission properties of the crystals were evaluated with a microscopic Fourier-transform infrared measurement system at a wide range of wavelengths (from 1.4 to 14 μm) at room temperature. The sample was characterized using a beam that is collimated within a 20° divergent angle, and the incident angles of the light were normal to and 20° from the vertical direction of the (100) surface of the crystals for the measurements of transmittance and reflectance, respectively. The area exposed to the infrared beam was fixed to the photonic pattern area using a slit. The reflection and transmission spectra of the crystals are shown in Fig. 3.8.9. Calculated predictions suggest that the fabricated crystals will have a stopband in the wavelength region of 3 – 5 μm. The crystals revealed a stopband in this expected wavelength region. In the measurements of reflection, a reflection peak at 3 – 4.5 μm became stronger and clearer as the number of layers increased. The intensity reached 75% with the eight-layered crystal. The amplitudes of reflectance peaks are less than those in other reports, whereas the transmittance spectra showed reasonable results.[1,2,18,19] This was due to the limitation of the incident angle of the reflectance measurement system. In the measurements of transmittance, a dip appeared in the same wavelength range where the reflectance peaks appeared, even with the four-layered crystal, and reached 3% with the eight-layered crystal. These measurements indicate that there is a significant PBG in the 3D crystals.

The technology discussed here enables the precise arrangement of objects thus allowing the realization of flexibility of design, a critical requirement in the fabrication of photonic crystal devices. To expand our technology further, approaches to enable production of larger crystals, introduction of a more complicated defect, and a combination of different materials are now underway. This technique will also evoke development of devices, which require the technology to introduce complicated defects, such as low-threshold lasers and photonic integrated circuits. A technology that can perform stable assembly in the order of a micron or submicron is indispensable not only to the field of photonic crystal devices but to the fields of electron devices and micromachining,[26] in which miniaturization is progressing at an increasing speed.

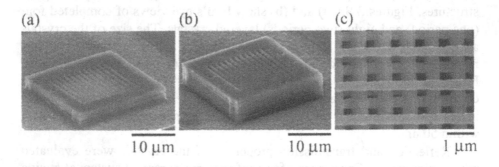

Figure 3.8.8 SEM images of 3D photonic crystals. (a) Four-layered stacked-stripe (or woodpile) structure. (b) Eight-layered stacked-stripe structure. (c) Surface view of the eight-layered crystal shown in (b).

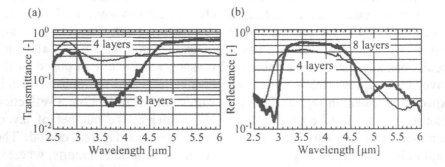

Figure 3.8.9 Optical properties of the obtained crystals. (a) Transmittance. (b) Reflectance.

References

1. S. Y. Lin, J. G. Fleming, D. L. Hetherington, B. K. Smith, R. Biswas, K. M. Ho, M. M. Sigalas, W. Zubrzycki, S. R. Kurtz and J. Bur, Nature **394**, 251 (1998).
2. S. Noda, K. Tomoda, N. Yamamoto, and A. Chutinan, Science **289**, 604 (2000).
3. H. Miguez, C. Lopez, F. Meseguer, A. Blanco, L. Vazquez, R. Mayoral, M. Ocana, V. Fornes, and A. Mifsud, Appl. Phys. Lett. **71**, 1148 (1997).
4. Particle Assembly Research Group, National Institute of Materials Science: The World of the Particle Integration Techniques — A new approach for devices and materials of the next generation, Kogyo-chosakai (2002, in Japanese)
5. H. Morishita and Y. Hatamura, Proc. IEEE/RSJ Int. Conf. Intelligent Robots and Systems, 1717 (1993).
6. H. Miyazaki and T. Sato, Advanced Robotics **11**, 169 (1997).
7. H. T. Miyazaki, Y. Tomizawa, S. Saito, T. Sato, and N. Shinya, J. Appl. Phys. **88**, 3330 (2000).
8. H. Tamaru, H. Kuwata, H. T. Miyazaki, and K. Miyano, Appl. Phys. Lett. **80**, 1826 (2002).
9. H. T. Miyazaki, H. Miyazaki, K. Ohtaka, and T. Sato, J. Appl. Phys. **87**, 7152 (2000).
10. K. M. Ho, C. T. Chan, and C. M. Soukoulis, Phys. Rev. Lett. **65**, 3152 (1990).
11. F. G. Santamaria, C. Lopez, F. Meseguer, F. Lopez-Tejeira, J. S. Dehesa, and H. T. Miyazaki, Appl. Phys. Lett. **79**, 2309 (2001).
12. F. G. Santamaria, H. T. Miyazaki, A. Urquia, M. Ibisate, M. Belmonte, N. Shinya, F. Meseguer, and C. Lopez, Adv. Mater. **14**, 1144 (2002) and the cover.
13. H. Hirayama, T. Hamano, and Y. Aoyagi, Appl. Phys. Lett. **69**, 791 (1996).
14. J. D. Joannopoulos, P. R. Villeneuve, and S. Fan, Nature **386**, 143 (1997).
15. C. C. Cheng and A. Scherer, J. Vac. Sci. Technol. B **13**, 2696 (1995).
16. J. E. G. J. Wijnhoven and W. L. Vos, Science **281**, 802 (1998).
17. A. A. Zakhidov, R. H. Baughman, Z. Iqbal, et al., Science **282**, 897 (1998).
18. A. Blanco, E. Chomski, S. Grabtchak, M. Ibisate, S. John, S. W. Leonard, C. Lopez, F. Meseguer, H. Miguez, J. P. Mondia, G. A. Ozin, O. Toader and H. M. van Driel, Nature **405**, 437 (2000).
19. Y. A. Vlasov, X. Z. Bo, J. C. Sturm, and D. J. Norris, Nature **414**, 289 (2001).
20. H. G. Sun, S. Matsuo and H. Misawa, Appl. Phys. Lett. **74**, 786 (1999).
21. M. Campbell, D. N. Sharp, M. T. Harrison, R. G. Denning, and A. J. Turberfield, Nature **404**, 53 (2000).·
22. S. Shoji and S. Kawata, Appl. Phys. Lett. **76**, 2668 (2000).
23. K. M. Ho, C. T. Chan, C. M. Soukoulis, R. Biswas, M. Sigalas, Solid State Commun. **89**, 413 (1994).
24. J. G. Fleming, S. Y. Lin, I. El-Kady, R. Biswas, and K. M. Ho, Nature **417**, 52 (2002).
25. K. Aoki, H. T. Miyazaki, H. Hirayama, K. Inoshita, T. Baba, N. Shinya and Y. Aoyagi, Appl. Phys. Lett. **81**, 3122 (2002).
26. D. Bishop, P. Gammel, and R. Giles, Physics Today **54**, 38 (2001).

(by H. T. Miyazaki and K. Aoki)

3.9 ORGANIC PHOTONIC CRYSTALS

3.9.1 Fabrication and characterization of 3D photonic crystals by self-organization of polymer microparticles

In this section, two examples of the fabrication methods of 3D photonic crystal by using organic material are explained. One of these is the self-organization of polymer microparticles[11] and the other is a 3D laser microfabrication method, which involves utilizing multiphoton absorption by the irradiation of a focused femtosecond laser pulse on an ultraviolet (UV)-curing resin and silica glass.[2-7] 3D photonic crystals made by these methods and their optical characterization are described in the following paragraph.

Recently, remarkable progress in the synthesis of polymer microparticles has made it possible to produce particles of a very uniform radius. Microparticles of polystyrene with a standard deviation in radius distribution of within 3% are already commercially available. 3D self-organization of 220 nm polystyrene microparticles can be made after water evaporation from a suspension, which is dropped into a microcell (with a size of $10 \times 10 \times 0.1$ mm) fabricated in the center of a coverslip using the lithography technique. An example of atomic force microscopy (AFM) image of the surface of self-organized 3D structure, which was fabricated without any control of the water evaporation velocity, is shown in Fig. 3.9.1(a).[11] It is clear from the figure that there are many defects in the organized structure of the microparticles at the surface. When the evaporation velocity is suppressed by humidity control, few defects are observed at the surface of the fabricated 3D structure of the microparticles, as is shown in Fig. 3.9.1(b).[11] The number of stacked layers in this 3D structure is estimated to be 40, according to the size of the microcell and the quantity of dropped solution. From surface observation by AFM, the 2D lattice of the microparticles was found to be triangular, that is, six microparticles lie around one particle and they come in to contact with each other. It is not clear which 3D filling arrangement is adopted by the structure because of a lack of cross-sectional images obtained by direct observation, but the 2D arrangement implies that either a hexagonal close-packed (hcp) or a cubic close-packed (ccp) crystal structure is present. This construction of such well-ordered 3D structures without defects is thought to be due to the slow and controlled evaporation of water after thermodynamic equilibrium is achieved, which results in the slow sedimentation of the particles. If the evaporation rate is faster, the movement of the microparticles is strongly suppressed due to the lack of water and the microparticles cannot settle and form a stable arrangement before the water completely evaporates, which leads to a defect-rich structure. The measurement of the transmission spectra from areas about 25 μm in diameter of both structures, displayed in Fig. 3.9.1, has shown that a PBG centered at 500 nm is observed, as is shown in Fig. 3.9.2. This indicates that the structure in Fig. 3.9.1(b) is very well self-organized with a high regularity in 3D.[11] However, the structure in

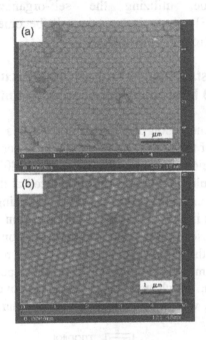

Figure 3.9.1 AFM surface images of the 3D photonic crystals made by self-organization of polystyrene microparticles.[12] The photonic crystals were fabricated (a) without, or (b) with, control of the water evaporation velocity.

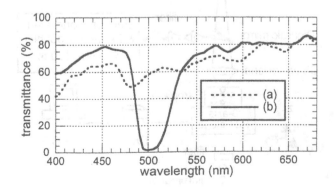

Figure 3.9.2 The transmission spectra of the 3D photonic crystals made by self-organization of polystyrene microparticles:[1] (a) and (b) correspond to those of the previous figure.

Fig. 3.9.1(a) seems to have numerous disordered defects in its 3D structure, since no such PBG was observed in the spectrum shown in Fig. 3.9.2.

The thermal adhesion of microparticles results in an increase in mechanical strength for this structure without a change in the transmission spectrum; the polystyrene particles that are in contact adhere firmly to each other after being kept at a high temperature for a certain time. Such a fabrication technique, utilizing the self-organization of polymer microparticles for a 3D photonic crystals, will become more important for the fabrication of micro photonic devices.

3.9.2 Photonic crystals made of an ultraviolet-curing resin utilizing a 3D laser microfabrication technique

The solidification phenomenon of UV-curing resin in a region of submicron-order in the vicinity of the laser focal point can be observed by the irradiation of focused femtosecond laser pulses (wavelength of 800 or 400 nm)[2-4]. This effect is due to the multi-photon absorption induced at the focus. Therefore it was tried to fabricate photonic crystals from resin utilizing this phenomenon.

The system used in this 3D laser microfabrication experiment is shown in Fig. 3.9.3.[4] The laser pulses used for multi-photon excitation were the second harmonic of the pulses from a Ti: sapphire regenerative amplifier (wavelength of 400 nm, pulse width of 120 fs, and repetition rate of 1 kHz). The UV-curing resin, which was sandwiched by coverslips and put on a triple axis PZT stage, was irradiated by the second harmonic pulses, which

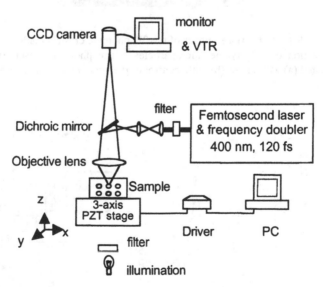

Figure 3.9.3 Experimental setup for 3D laser microfabrication.[4]

were focused to nearly the diffraction limit, using an oil immersion objective lens (×100, numerical aperture of 0.8 – 1.3). The fabrication process of the 3D structure was observed *in-situ* using a CCD camera and was recorded by

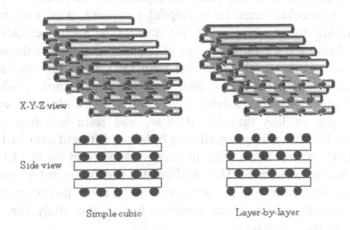

Figure 3.9.4 Two examples of 3D photonic crystal structures. Simple cubic (left), and stacked-stripe (right). The upper panels show *x-y-z* views and the lower panels show side views.

Figure 3.9.5 Stacked-stripe photonic crystal fabricated by 3D laser microfabrication technique.[2] (a) Top view (optical microscope image). (b) cross-section (SEM image).

a video recorder. Nopco-cure 800, which does not absorb light with a wavelength longer than 370 nm, was used as the UV-curing resin for this experiment.

The structures of the 3D photonic crystals shown in Fig. 3.9.4, which are called the stacked-stripe (or woodpile) structures, were each fabricated by controlling the movement of the three-axis Piezo actuator with a computer. Since the wavelength range of the PBG depends on the width and period of the submicrometer rods that form the crystal, the irradiated laser power was changed so as to determine the rod diameter, which was measured afterwards. The solidified product was not observed with laser pulse energies of less than 80 nJ/pulse, and resin bubbling due to a temperature increase was observed with light intensities of over 140 nJ/pulse. Consequently, a laser beam with an energy range from 80 to 140 nJ/pulse was used in this experiment. The rod diameter was controlled by varying the irradiating laser power. The diameter can also be controlled by changing the scanning velocity of the Piezo actuator, but in this study the scanning velocity was fixed at 10 μm/s.

The top view of the fabricated photonic crystal obtained by optical microscopy and its cross-section by SEM are shown in Fig. 3.9.5. From the SEM image, it is clear that the rod cross-section is almost a circle. The SEM image also shows that the rods of the second nearest layers are shifted by one-half period in the direction perpendicular to the rod axes, which is

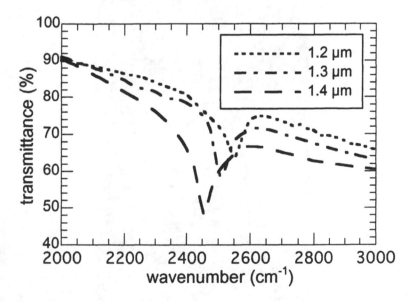

Figure 3.9.6 Transmission spectra of the fabricated stacked-stripe structure photonic crystals.[2] The lattice period is shown in Fig. 3.9.5.

required for the stacked-stripe structure.

The transmission spectrum of the fabricated photonic crystal (20 layers), which was measured by FT-IR, is shown in Fig. 3.9.6. All rods for the structure were fabricated at a laser power of 90 nJ/pulse and their diameters are 1.0 μm. It was confirmed that as the period of the rods increases, to 1.2, 1.3, and 1.4 μm, the peak of the PBG shifts to longer wavelength (smaller wavenumber) as is demonstrated by the peaks at 2553, 2507, and 2454 cm^{-1}. These results indicate that the observed peaks in the transmission spectrum are due to the PBG. There are two possible reasons for the small change in observed transmittance, which is only a few tens of percent:

(i) The refractive index of the solidified product obtained from the UV-curing resin is not large (it is only about 1.6).

(ii) Rod diameter fluctuated slightly because of instability in the incident laser power.

At present, in order to increase the refractive index of this photonic crystal, several different strategies are being pursued, such as the addition of a higher refractive index material to the resin and a technique utilizing a sol-gel reaction whereby a resin-based photonic crystal is used as a mold for a photonic crystal produced from a material with a higher refractive index. Recently, a dye with a large cross-section for two photon absorption has been developed,[8] which will enable this method to be used to fabricate a 3D photonic crystal with lower laser power than in conventional methods. Therefore, this type of method is expected to be important for future photonic crystal research.

References

1. K. Fukuda, H. -B. Sun, S. Matsuo, and H. Misawa, Jpn. J. Appl. Phys. 37, L508 (1998).
2. H. -B. Sun, S. Matsuo, and H. Misawa, Appl. Phys. Lett. 74, 786 (1999).
3. H. -B. Sun, Y. Xu, S. Matsuo, and H. Misawa, Opt. Rev. 6, 396 (1999).
4. M. Watanabe, H. -B. Sun, S. Juodkazis, T. Takahashi, S. Matsuo, Y. Suzuki, J. Nishii, and H. Misawa, Jpn. J. Appl. Phys. 37, L1527 (1998).
5. M. Watanabe, S. Juodkazis, H.-B. Sun, S. Matsuo, H. Misawa, M. Miwa, and R. Kaneko, Appl. Phys. Lett. 74, 3957 (1999).
6. M. Watanabe, S. Juodkazis, H. -B. Sun, S. Matsuo, and H. Misawa, Phys. Rev. B 60, 9959 (1999).
7. H. Misawa, J. Spectroscopical Soc. Jpn. (Bunkokenkyu) 48, 74 (1999, in Japanese).
8. B. H. Cumpston, S. P. Anaanthavel, S. Barlow, D. L. Dyer, J. E. Ehrlich, L. L. Erskine, A. A. Heikal, S. M. Kuebler, I. -Y. S. Lee, D. McCord-Maughon, J. Qin, H. Röckel, M. Rumi, X.-L. Wu, S. R. Marder, and J. W. Perry, Nature 398, 51 (1999).

(by H. Misawa)

3.10 PHOTONIC CRYSTALS IN TERAHERTZ REGION

3.10.1 General features

The intermediate region between the visible and radio waves, comprising the frequency range from about 0.1 THz to 10 THz, is known as the terahertz (THz) region. This frequency range covers the spectrum from millimeter waves to infrared waves. Although optical technology has progressed in the visible wave region and the technology for electric and electronic communications has also progressed in the radio wave region, the THz region has fallen behind the progress of these developments with regard to almost all elemental technologies such as sources (oscillators), detectors (receivers), and spectroscopic methods. Mobile communications, especially portable phones, require operative devices in the THz frequency range in order to increase the communication capacity. As a result, research into elemental technologies in the THz region is urgently needed to develop devices such as a monochromatic wave source with a higher efficiency and a detector with higher accuracy. Technological developments in the THz electromagnetic wave range will become increasingly important because of the stimulation of technical progress in the peripheral region and an increasing demand in the application field. The integrated development of new technology is indispensable, with regards to the generation, the translation, the modulation, and the detection of electromagnetic waves if the THz wave range is to be utilized for communications.

As the scaling law between the lattice constant and the wavelength of an electromagnetic wave in photonic crystals is well established, it is possible to develop high efficiency devices for the necessary wavelength range by applying elemental techniques, such as lattice fabrication or spectroscopy. In particular, the items that will be developed are likely to include a high Q-value resonator for high efficiency oscillations, a low loss waveguide and a narrow band filter utilizing the impurity localization mode in a photonic crystal.[1-4] At present, photonic crystal applications for the THz region are limited to passive devices, but if the techniques to introduce oscillators or emission materials into the photonic crystal lattice are developed, active devices can be realized as well.

The first studies on photonic crystals were concentrated in the microwave range because such crystals are easy to prepare. The vector of studies is now directed to the near infrared and visible regions because of the importance of controlling the spontaneous emission in pure physics and of the importance of developing applications in laser oscillation or nonlinear optical effects. However, developments in the THz region are not merely a stage in the progression from longer to shorter wavelengths. Rather, considering the development of practical application techniques, great opportunities lie in the THz region, including the development of millimeter wave devices. The purpose of a photonic crystal study in the THz region is the direct acquisition of knowledge concerning the propagation characteristics of electromagnetic waves, such as the dispersion relation of electromagnetic waves in a photonic crystal, which is difficult to measure directly by using any techniques in the infrared or visible regions, as well as device development in this wavelength region. Attention has now turned to the effectiveness with which the knowledge obtained in the THz region can be applied to other wavelength regions, such as the infrared or the visible region, as the scaling law is well established between the lattice constant and the wavelength of an electromagnetic wave in a photonic crystal. THz time-domain spectroscopy (THz-TDS) using a pulsed THz electromagnetic wave has also been established since the development of a detection and generation technique for pulsed THz electromagnetic waves utilizing an ultrashort pulsed laser.[5-6] This is based on the recent rapid progress made in femtosecond lasers. The fabrication methods for photonic crystals in the THz region and the studies using THz-TDS regarding the propagation characteristics of THz waves in photonic crystals are introduced in this chapter.

3.10.2 Fabrication of terahertz range photonic crystals

Although the techniques used to fabricate photonic crystals for the infrared and visible regions have progressed dramatically for 1D and 2D crystals, 3D crystal fabrication remains complex. However, fabrication of THz region photonic crystals is simpler than that of infrared and visible region photonic crystals because the lattice constant is of the order from millimeters to sub-millimeters. In this region it is relatively easy to fabricate even 3D crystals with a high accuracy. Almost all of the fabrication techniques used in the shorter wavelength range such as building blocks, wafer fusion, stereolithography, hole opening processes, and sphere arrangement techniques can be applied to the fabrication of terahertz range photonic crystals.

The stacked-stripe photonic crystal in the microwave region was fabricated by stacking Si plates processed by a Si double etching method, and subsequently realized a PBG with a center frequency of 350 GHz in the millimeter wave region.[7] The single cubic lattice, where dielectric rods running parallel to the x, y, and z axes gather at one point also has a PBG opened regardless of the rod figure being a circular pillar, a square dielectric pillar or even an air rod.[8] This structure attracted researchers' attention because it can be fabricated by stacking plates on which surfaces holes and caves are opened. This photonic crystal structure is expected to be used as an active optical device in the THz region, especially because it is easy to process mechanically to a submillimeter order and as it is relatively easy to introduce a defect layer as an impurity by means of the layer stacking process itself and to thereby incorporate an oscillating device.

The *wafer fusion* technique, which produces one of the basic building block structures, is to repeat the combination of the previously combined objects so as to create a stack,[9] as described in Section 3.6. For mass-production of a multi-layer-stacked crystal, a fine 3D shape generation method to translate and stack the film patterns, which are formed as a lump on each sectional structure of a 3D structure by photolithography on the target substrate using a room temperature bonding technique, has been proposed.[10]

A rapid prototyping method known as *stereolithography* is used to form an arbitrary crystal structure, which uses ultraviolet ray hardened epoxy or two-photon absorbing polymerized mediums. A complete diamond structure crystal was realized by means of stereolithography.[11] In this structure, the four carbon atom covalent bonds in real diamond crystal are replaced by dielectric rods, in which a relatively wide common PBG is expected. The opposite, of this structure, an inverse diamond structure, in which the covalent bonds are replaced by air rods, has also been fabricated. Both of these structures were confirmed to have a wide common PBG experimentally. The fabrication method using two-photon absorbing polymerized mediums has been investigated,[12] as described in Section 3.9. The main characteristics of the optical molding method are its integrated fabrication method for an electromagnetic waveguide and oscillator, as well as for a photonic crystal lattice, by assigning the necessary space required for each device at the design stage.

Hole opening is the simplest technique for producing pseudo-diamond structures. A PBG was demonstrated in a microwave region crystal, which was fabricated by opening holes mechanically from three directions in a material with a high dielectric constant.[2] Afterwards, the propagating characteristics of electromagnetic waves in photonic crystals of a single cubic lattice of Ge or triangular and square lattices of a polymer (TPX) were analyzed using the THz-TDS.[13-14]

Ordering of spheres is relatively easy once spheres that are well matched in shape and size have been prepared. In close-packed face centered cubic lattice photonic crystals, a PBG only exists when the dielectric spheres are slightly superimposed. Conversely, a relatively wide PBG exists in the inverse close-packed fcc lattice (inverse opal) structure, in which the spheres and the remainder of the lattice are replaced by air and dielectrics, respectively. A positive-type crystal is first fabricated by use of polymer spheres such as polystyrene, and used as a replica to fabricate a negative-type photonic crystal.[15] This method is a potential candidate for the manufacturing process used for photonic crystal devices from the microwave to visible regions, because it is easy to prepare the well matched polystyrene spheres of the size required for these wavelengths, though this method demands some further innovations in device design. Potential improvements include the integration of a defect layer and an emission device. In the THz region, the effect of the dielectric loss on the transmission spectrum was considered using a crystal, which was fabricated by stacking layers with a triangular lattice arrangement of Si_3N_4 dielectric spheres.[16]

Textile is a method, which enables photonic crystals to be fabricated with dielectric threads, in a manner resembling the weaving of cloth, is unique. This technique was established in order to fabricate a stacked-stripe structure by using 3D fiber textile and confirmed that their sample possessed a wide PBG.[17]

3.10.3 Determination of electromagnetic dispersion by terahertz time-domain spectroscopy

It is vital to consider the propagating characteristics of an electromagnetic wave in a photonic crystal from not only a theoretical, but also an experimental viewpoint when designing optical devices using photonic crystals, because the sizes of the optical devices in actual use are finite. If a subjective photonic crystal lattice for the THz region is fabricated in order to estimate the dispersion relation by using THz-TDS, knowledge necessary for the development of devices operating in the invisible region is obtained via the scaling law.

Time-domain spectroscopy is a method to obtain the light spectrum by using a time sequential Fourier transformation of the measured time dependency of changes in the electric field of electromagnetic waves. Its main characteristic is that the amplitude of the electric field and phase shift spectra are obtained both independently and simultaneously. Since the phase shift is proportional to the absolute value of the wavevector of the electromagnetic wave, the dispersion relation $k(\omega)$ of the electromagnetic wave in a medium can be determined directly from the phase shift spectra, $\phi(\omega)$.[6] A schematic of the equipment used in THz-TDS is shown in Fig. 3.10.1. It consists of an ultrashort pulsed laser (a mode-locked Ti-sapphire laser or an Er-doped fiber laser), an emitter (a photo-conductive antenna or a nonlinear optical crystal), an optical system (such as a Si hemispherical lens and pairs of off-axis parabolic mirrors), a photodetector (a photo-conductive antenna or an EO crystal), a time delay circuit, and a workstation for controlling the system and performing the analysis.

The transmission intensity and phase shift spectra in the Γ-Z direction

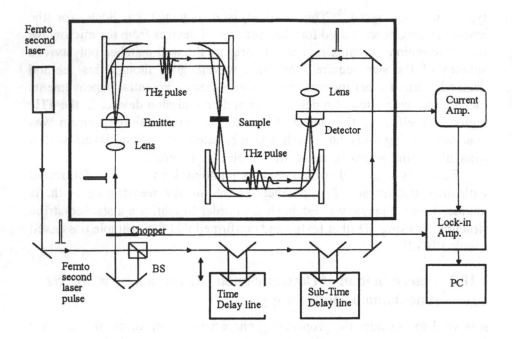

Figure 3.10.1 The schematic of THz time-domain spectroscopy (THz-TDS).

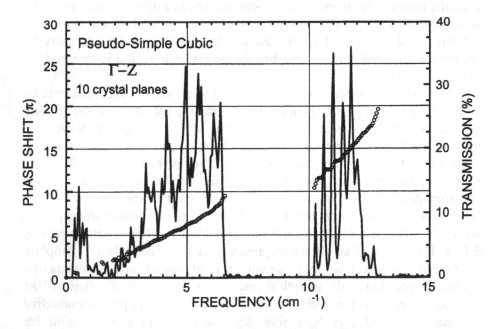

Figure 3.10.2 The transmission intensity and phase shift spectra in the Γ-Z direction of a pseudo-single cubic air rod lattice photonic crystal made of Si.

of a pseudo-single cubic air rod lattice (lattice constant of 0.4 mm and 10 unit cells) made of Si are shown in Fig. 3.10.2. An opaque region due to the first PBG appears from 6.5 cm^{-1} to 10.2 cm^{-1}. The phase shifts by 10π at the center of the first PBG. It corresponds to the 10 passes of an electromagnetic wave in the Γ-Z direction through the (001) crystal plane. The modulation caused by the Fabry-Perot interference effect between the incident and radiating surfaces also appears in the phase shift spectrum. The maximum gradient of phase shift corresponds to the maximum frequency of the transmission intensity and these also agree with the maximum optical density of state of standing waves.[18] Between the absolute value of wavevector $k(\omega)$ and the phase shift $\phi(\omega)$ for an electromagnetic wave, there is a simple proportional relation,

$$k(\omega) = \phi(\omega)/d - \omega/c, \tag{1}$$

where d is the thickness of the sample in the transmission direction, and c is the light velocity. The dispersion relation of an electromagnetic wave in a photonic crystal can be calculated with the above equation. The dispersion relations of THz waves propagating to the Γ-Z and Γ-M directions of a pseudo-single cubic air-rod lattice, which were calculated from the phase shift spectra, are shown in Fig. 3.10.3. The dispersion relation of an H-polarized electromagnetic wave in a square lattice photonic crystal, which is made of TPX with air-rods, is shown in Fig. 3.10.4. A solid line and a white circle indicate the calculated results from the plane wave expansion method and the measured results estimated from the phase shift spectra, respectively. Both show very good coincidence with each other, not only qualitatively, but also quantitatively.[19] It should be noted that the almost zero group velocity of the electromagnetic wave at the Brillouin zone center and boundaries can be directly confirmed by using THz-TDS.

References

1. K. Ohtaka, Phys. Rev. B **19**, 5057 (1979).
2. E. Yablonovich, Phys Rev. Lett. **58**, 2059 (1987).
3. For a review, see Photonic Band Gaps and Localization (Ed. by C. M. Soukoulis), Plenum Press (1993), and special review issues in J. Opt. Soc. Am. **B10**, (1993).
4. K. Sakoda, Optical Properties of Photonic Crystals, Springer Verlag, (2001).
5. see Special review issues on Terahertz electromagnetic pulse generation, physics, and applications, J. Opt. Soc. Am. **B11**, 2454 (1994).
6. M. Tani, S. Matsuura, K. Sakai, and S. Nakashima, Appl. Opt. **36**, 7853 (1997).
7. E. Ozbay, A. Abeyta, G. Tuttle, M. Tringides, R. Biswas, C. T. Chan, C. M. Soukoulis, and K. M. Ho, Phys. Rev. B **50**, 1945 (1994); E. Ozbay, E. Michel, G. Tuttle, R. Biswas, K. M. Ho, J. Bostak, and D. M. Bloom, Appl. Phys. Lett. **65**, 1617 (1994).
8. J. W. Haus, J. Mod. Opt. **41**, 195 (1994).
9. S. Noda, K. Tomoda, N. Yamamoto, and A. Chuitan, Science **289**, 604 (2000).

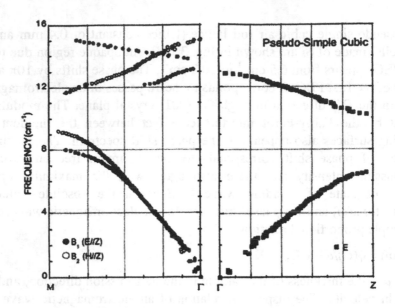

Figure 3.10.3 The dispersion relation of an electromagnetic wave in a pseudo-single cubic air rod lattice photonic crystal made of Si.

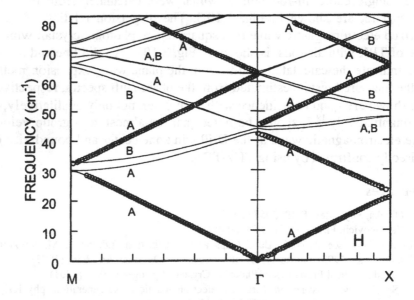

Figure 3.10.4 The dispersion relation of an *H*-polarized THz wave in a square lattice photonic crystal made of TPX with air-rods. The solid line and the open circles denote the results calculated by the plane wave expansion method and the measured results estimated from the phase shift spectra, respectively.

10. T. Yamada and R. Takahashi, Jpn. Soc. Precision Eng. (Seimitsukogaku) **66**, 1265 (2000, in Japanese).
11. S. Kirihara, Y. Miyamoto, K. Takenaga, M. Wada Takeda, and K. Kajiyama, Solid State Commun. **121**, 435 (2002).
12. T. Kondo, S. Matsuo, S. Juodkaris, and H. Misawa, Appl. Phys. Lett. **79**, 725 (2001).
13. M. Wada, Y. Doi, K. Inoue, J. W. Haus, and Z. Yuan, Appl. Phys. Lett. **70**, 2966 (1997).
14. M. Wada, Y. Doi, K. Inoue, and J. W. Haus, Phys. Rev. B **55**, 10443 (1997).
15. S. G. Romanov, T. Make, C. M. Sotomayor Torres, M. Muller, and R. Zentel, Appl. Phys. Lett. **75**, No. 8 (1999).
16. S. Yamaguchi, S. Yano, J. Inoue, K. Ohtaka, and Y. Segawa, Proc. Annual Meet. Jpn. Phys. Soc. **57-1**, 24aWG-5 (2002, in Japanese).
17. Y. Watanabe, T. Kobyashi, S. Kirihara, and Y. Miyamoto, Proc. Asian Textile Conf. **252**, 1-8 (2001).
18. T. Aoki, M. Wada Takeda, J. W. Haus, Z. Yuan, M. Tani, K. Sakai, N. Kawai, and K. Inoue, Phys. Rev. B **64**, 45106 (2001).
19. H. Kitahara, N. Tsumura, H. Kondo, M. Wada Takeda, J. W. Haus, Z. Yuan, N. Kawai, K. Sakoda, and K. Inoue, Phys. Rev. B **64**, 45202 (2001).

(by M. Takeda)

3.11 PHOTONIC CRYSTAL FIBERS

3.11.1 Overview

Photonic crystal fiber (PCF) is an optical fiber that contains photonic crystal structures. It is usually realized by periodically arranging airholes in silica glass, which extend along the fiber. The most important features are that it can guide light via a PBG and that we can expect novel functions, such as ultralow loss and ultralow nonlinearity transmission over a hollow core, which cannot be obtained with conventional optical fibers based on total reflection guidance, to be realized. Moreover, even when they are based on total reflection guidance, optical fibers with airholes can provide various new characteristics such as a large waveguide dispersion and single mode operation over a wide wavelength range. Optical fiber with airholes and total reflection based operation is called holey fiber or microstructured fiber. It is also a feature of holey fibers that they can be made of a single material, which expands the selection range of materials for optical fibers. Since the first proposal,[1] photonic crystal fibers and holey fibers have been intensively researched, and recently even a transmission experiment was carried out.[2] In this report, the results of this research are described with a focus on fabrication methods.

3.11.2 Photonic crystal fibers

Examples of the cross-sectional structures of PCF are shown in Fig. 3.11.1. A photonic crystal structure is realized in each example by arranging airholes in silica glass periodically. If the structure is designed appropriately, there exists a range, called the PBG, of light frequency in which light propagation in the photonic crystal is forbidden. When a defect disturbing the periodicity is introduced into a photonic crystal having a PBG, light of frequencies corresponding to the PBG becomes localized at the defect. Since the photonic crystal and the defect in the PCF extend along the direction of the fiber axis, the light propagates along the defect through the fiber. Thus, the defect and the photonic crystal function as the core and the cladding, respectively. The PBG in the PCF has a feature that the wavevector component normal to the plane of periodicity helps the PBG to open even if the refractive index contrast is low.[1] As a result, a PBG can be observed even with a low refractive index contrast of about 1.45 between silica glass and air.

Figure 3.11.1(a) shows the PCF structure that was first produced.[3] The airholes arranged in a triangular lattice form a photonic crystal in the cladding, and a defect is made in the core by replacing an airhole with silica

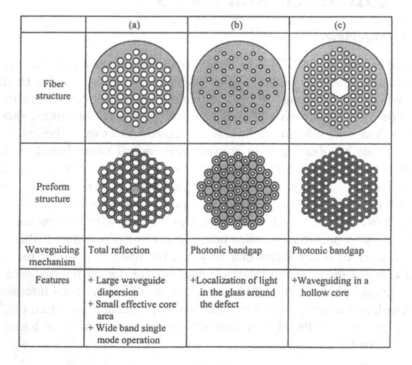

	(a)	(b)	(c)
Fiber structure			
Preform structure			
Waveguiding mechanism	Total reflection	Photonic bandgap	Photonic bandgap
Features	+ Large waveguide dispersion + Small effective core area + Wide band single mode operation	+Localization of light in the glass around the defect	+Waveguiding in a hollow core

Figure 3.11.1 The structures and features of photonic crystal fibers.

glass. The diameter and interval pitch of the airholes were 0.35 μm and 2.3 μm, respectively. Since the average refractive index at the defect is higher than that in the photonic crystal, this structure operates via total internal reflection. Thus, the structure shown in Fig. 3.11.1(a) is also classified as holey fiber. Although modes guided by a PBG can, in theory, exist separately from the modes guided by total reflection,[4] they have not yet been observed. A structure like that in Fig. 3.11.1(a) can be fabricated by a method in which glass capillaries and glass rods are bundled to make a preform and a fiber is then drawn from this preform in a furnace at a temperature of about 2000 °C, as shown in Fig. 3.11.2. The typical diameter of the capillaries and the rods is about 1 mm, and the cross-sectional structure is reduced by a factor of 1/500 in the fiber drawing process, so that a structure with an airhole pitch of about 2 μm can be obtained. In this fabrication method, a step to insert the bundle of capillaries into a glass tube is often included and the bundle and the tube are then merged in the drawing step, so that the fiber diameter can be increased and the mechanical strength can be improved. In the drawing step, a fine tuning of drawing conditions is important to prevent the collapse of the airholes due to surface tension. This method, called the multiple-capillary method, is the standard method for fabricating PCF's and holey fibers. Whilst capillaries and a rod with hexagonal pillar shapes, as in Fig. 3.11.1(a), were used in the first PCF

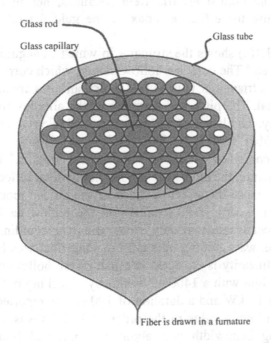

Figure 3.11.2 A schematic of drawing of a photonic crystal fiber.

fabrication, similar fiber structures were later realized using capillaries and rods having a cylindrical shapes.[5] However, when capillaries and rods having cylindrical shapes are employed, interstitial holes between three adjacent capillaries or rods often remain in the fiber. These interstitial holes do not often affect the characteristics of the holey fiber, but in a structure based on a PBG, like that in Fig. 3.11.1(b), they can increase the width of the PBG.[6] The fabrication of photonic crystals by fiber drawing has the advantages that a structure with a dimension of the order of the wavelength of the light can be obtained from a structure with a dimension of the order of several millimeters, which is easy to process mechanically, and that a waveguide structure with a length of the order of several kilometers can be obtained.

Figure 3.11.1(b) shows the structure in which waveguiding by PBG was first observed.[7] The airhole arrangement, known as a honeycomb structure, realizes the photonic crystal in the cladding. The diameter and pitch of the airholes are 0.8 μm and 1.9 μm, respectively. The preform was fabricated by combining capillaries and rods. Guidance of light in the wavelength range of 458 – 528 nm was observed in a 5 cm fiber. Since the defect in the core is realized by introducing an excess airhole, the average refractive index of the core defect is lower than that of the cladding so that no modes can be guided by total reflection. Therefore, the guided modes are based on the PBG. In this structure, the optical electric field localizes, not in the airhole, but around it, because the effective index of the guided mode is higher than unity.

Figure 3.11.1(c) shows the structure in which waveguiding in a hollow core was observed.[8] The core is a hollow cavity, which corresponds to seven lattice points of a triangular lattice upon which airholes are arranged to form a photonic crystal. The pitch and diameter of the airholes are 4.9 μm and 3 μm, respectively. Guidance of light in the visible wavelength range was observed in a fiber with 3 cm long. Hollow cores may be formed by a technique that combines capillaries of different lengths.[9] Because of the hollow core structure, the modes are guided not by total reflection, but by the PBG. Moreover, since the effective index of the guided mode is lower than unity, the optical electric field localizes in the hollow cavity of the core. Although the fiber length was only 3 cm when waveguiding was observed for the first time, waveguiding over a 15 m long fiber was later observed.[10] Whilst low nonlinearity is expected because of the hollow core, generation of a broadband light with a 1400 nm width by launching optical pulses with a peak power of 11 kW and a duration of 100 fs was reported.[11] Concerning the transmission loss, a value of the order of 1 dB/m was reported.[12] Here, the waveguiding bandwidth was about 100 nm and both positive and negative chromatic dispersion, having absolute values of 500 – 1000 ps/nm·km, were observed in the band-edges. PBG-guided fibers have made the most remarkable progression of all PCF's in recent years, and further investigations are expected in attempts to improve their characteristics and applications.

3.11.3 Holey fibers

Holey fiber is an optical fiber in which the core-cladding structure necessary for waveguiding by total reflection is effectively realized by the arrangement of airholes. Whilst photonic crystal fibers based on total reflection can be classified as holey fibers, such periodical structures as those found in photonic crystals are not required in holey fibers. Holey fiber has the following features. Refractive index contrast in the cross-section is higher than that in conventional optical fibers, distribution of refractive index in the cross-section is fully 2D and fabrication from a single material is possible. As a result, novel functions and characteristics not available in conventional optical fibers can be realized, and in addition, they can be more easily fabricated than the PBG-guided fibers. Hence, many groups have intensively researched holey fibers. The typical method of fabrication is the multiple-capillary method similar to that used for photonic crystal fibers.

The first feature of holey fiber is its refractive index contrast. Whilst the refractive index contrast in the cross-section has been at most several percent in conventional optical fibers, the relative refractive index difference between silica glass and air is as large as about 26%. As a result, large waveguide dispersion, small effective core area, and high birefringence can be realized. The large waveguide dispersion can realize chromatic dispersion characteristics that have been almost impossible previously. These are, for example, (1) a large negative chromatic dispersion of about −2000 ps/nm·km in the 1.55 μm band,[13] (2) a zero or an anomalous dispersion in the wavelength range shorter than 1.28 μm,[14,15] (3) a flat (hardly wavelength-dependent) chromatic dispersion in the 1.55 μm band.[16] Possible applications of these characteristics are considered to be as follows: (1) highly efficient dispersion compensation for the standard single mode fiber, (2) utilization of nonlinear optical effects in the short wavelength range, such as soliton transmission, generation of supercontinuum light, and optical parametric amplification,[17] and (3) a large-capacity transmission medium. Among these, many application experiments regarding (2) have been reported. Also, small effective core areas, of $2 - 3 \ \mu m^2$, have become available in the 1.55 μm wavelength band, which enables highly efficient generation of nonlinear optical effects even with a short fiber length. Exploiting these features, many applications have been demonstrated, such as pulsed light sources,[18,19] an optical switch,[20] a Raman amplifier,[21] an optical CDMA receiver,[22] and so on.

Moreover, regarding birefringence, a high birefringence of 1.4×10^{-3}, which is several times higher than that of conventional polarization-maintaining fibers, was realized.[2]

The second feature of holey fiber is that it has a 2D refractive index distribution with a scale of wavelength order. As a result, the effective refractive index, which is defined here as the refractive index of a uniform medium equivalent to the 2D refractive index distribution, becomes strongly dependent on wavelength and the scale of the structure.[23] This feature resulted in realization of wide-band single mode operation over a range of at least $337 - 1550$ nm[24] and a large effective core area of several hundred square micrometers.[24] The former is desirable for wide-band transmission, and the latter is desirable for suppressing unwanted nonlinear optical effects.

The third feature of holey fiber is that it can be fabricated from a single material. Whilst fabrication of conventional optical fiber requires at least two kinds of media with different refractive indices, a single material is sufficient for holey fiber in which the waveguiding structure is realized by airholes, so that the selection range for the material becomes wider than before. In addition to those using silica glass, fabrications using a compound glass[25] and a polymer[26] have been reported.

For practical applications of holey fibers, it is important to reduce the transmission loss and the connection loss. Whilst the transmission loss of conventional optical fibers is $0.2 - 1$ dB/km in the 1.55 μm band, transmission losses in holey fibers in the early stage of their development were as high as several tens of dB/km.[27] However, the fabrication technique has been improved recently, and reports of transmission losses of $1 - 3$ dB/km[12,28,29] have been published. Also, a low loss of 0.4 dB/km is realized by the hole-assisted lightguide fiber, which is a combination of airholes and a refractive index distribution similar to that of conventional optical fibers.[30] Conversely, the input and output losses of the fiber may be a problem, especially in holey fibers with small core diameters designed for nonlinearity utilization. Although there has been a proposal to realize both connectibility and high nonlinearity by tapering a holey fiber with a large core diameter,[18] the applicable fiber length is still short. Investigation into other techniques of connection would be necessary.

3.11.4 Bragg fiber

Bragg fiber is an optical fiber similar in structure to the PBG fibers. As shown in Fig. 3.11.3, it has a cross-sectional refractive index profile periodical in the radial direction. This periodical distribution of the refractive index thus works as a Bragg mirror and guides light along the fiber by Bragg reflection. Whilst the confinement of light to a hollow core was proposed,[31] it is known to be difficult to transmit it over a fiber with a low leakage loss because the refractive index difference available in silica glass is small.[32] However, a Bragg fiber has recently been fabricated from tellurium (a refractive index of 4.6) and polystyrene (1.59). [33] Although the theory predicts that the leakage loss can be suppressed to an order of 10^{-3} dB/km by stacking 17 layers, it has not yet been confirmed experimentally. Since the fabrication method is not fiber-drawing but consists of direct deposition of multilayer on the fiber, there is a problem that the fiber length is as short as 10 cm. Nonetheless, once appropriate materials are found, its application to transmission mediums with ultralow loss and ultralow nonlinearity would be realized, because it is expected that the contribution of the material loss of the Bragg mirror to the transmission loss can be reduced to about 10^{-3}.

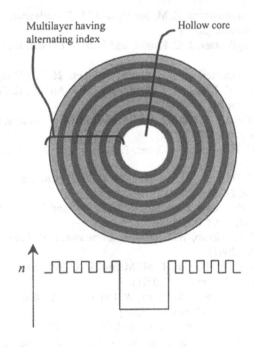

Figure 3.11.3 The cross-sectional structure and the refractive index distribution of a Bragg fiber.

References

1. T. A. Birks, P. J. Roberts, P. St. J. Russell, D. M. Atkin, and T. J. Shepherd, Electron. Lett. **31**, 1941 (1995).
2. K. Suzuki, H. Kubota, S. Kawanishi, M. Tanaka, and M. Fujita, Electron. Lett. **37**, 1399 (2001).
3. J. C. Knight, T. A. Birks, P. St. J. Russell, and D. M. Atkin, Opt. Lett. **21**, 1547 (1996).

4. A. Ferrando, E. Silvestre, J. J. Miret, P. Andres, and M. V. Andres, Opt. Lett. **25**, 1328 (2001).
5. J. C. Knight, T. A. Birks, R. F. Cregan, P. St. J. Russell, and J. -P. de Sandro, Opt. Mat. **11**, 143 (1999).
6. J. Broeng, S. E. Barkou, A. Bjarklev, J. C. Knight, T. A. Birks, and P. St. J. Russell, Opt. Comm. **156**, 240 (1998).
7. J. C. Knight, J. Broeng, T. A. Birks, and P. St. J. Russell, Science **282**, 1476 (1998).
8. R. F. Cregan, B. J. Mangan, J. C. Knight, T. A. Birks, P. St. J. Russell, P. J. Roberts, and D. C. Allan, Science **285**, 1537 (1999).
9. P. St. J. Russell, S. Mead, T. A. Birks, J. C. Knight, and B. J. Mangan, Int. Patent Appl., WO00/60388 (2000).
10. J. A. West, J. C. Fajardo, M. T. Gallagher, K. W. Koch, N. F. Borrelli, and D. C. Allan, Proc. European Conf. Opt. Commun., 41 (2000).
11. K. Suzuki, and M. Nakazawa, Proc. Optoelectronic and Commun. Conf., WIPD1-11 (2001).
12. J. A. West, N. Venkataramam, C. M. Smith, and M. T. Gallagher, Proc. European Conf. Opt. Commun., Th.A.2.2 (2001).
13. T. A. Birks, D. Mogilevtsev, J. C. Knight, and P. St. J. Russell, IEEE Photon. Tech. Lett. **11**, 674 (1999).
14. J. K. Ranka, R. S. Windeler, and A. J. Stentz, Opt. Lett. **25**, 25 (2000).
15. W. J. Wadsworth, J. C. Knight, A. Ortigosa-Blanch, J. Arriaga, E. Silvestre, and P. St. J. Russell, Electron. Lett. **36**, 53 (2000).
16. A. Ferrando, E. Silvestre, J. J. Miret, J. A. Monsoriu, M. V. Andres, and P. St. J. Russell, Electron. Lett. **35**, 325 (1999).
17. J. E. Sharping, M. Fiorentino, A. Coker, P. Kumar, and R. S. Windeler, Opt. Lett. **26**, 1048 (2001).
18. X. Liu, C. Xu, W. H. Knox, J. K. Chandalia, B. J. Eggleton, S. G. Kosinski, and R. S. Windeler, Opt. Lett. **26**, 358 (2001).
19. J. H. V. Price, K. Furusawa, T. M. Monro, L. Lefort, and D. J. Richardson, Proc. Conf. Laser and Electro-Optics, CPD1 (2001).
20. P. Petropoulos, T. M. Monro, W. Belardi, K. Furusawa, J. H. Lee, and D. J. Richardson, Opt. Lett. **26**, 1233 (2001).
21. J. H. Lee, Z. Yusoff, W. Belardi, T. M. Monro, P. C. Teh, and D. J. Richardson, Proc. European Conf. Opt. Commun., 46 (2001).
22. J. H. Lee, P. C. Teh, Z. Yusoff, M. Ibsen, W. Belardi, T. M. Monro, and D. J. Richardson, Proc. European Conf. Opt. Commun., (2001).
23. T. A. Birks, J. C. Knight, and P. St. J. Russell, Opt. Lett. **22**, 961 (1997).
24. J. C. Baggett, T. M. Monro, K. Furusawa, and D. J. Richardson, Opt. Lett. **26**, 1045 (2001).
25. T. M. Monro, Y. D. West, D. W. Hewak, N. G. R. Broderick, and D. J. Richardson, Electron. Lett. **36**, 1998 (2000).
26. M. A. van Eijkelenborg, M. C. J. Large, A. Argyros, J. Zagari, S. Manos, N. A. Issa, I. Bassett, S. Fleming, R. C. McPhedran, C. M. de Sterke, and N. A. P. Nicorovici, Opt. Exp. **9**, 319 (2001).
27. D. C. Allan, N. F. Borrelli, J. C. Fajardo, R. M. Fiacco, D. W. Hawtof, and J. A. West, Int. Patent Appl., WO 00/37974 (2000).
28. H. Kubota, K. Suzuki, S. Kawanishi, M. Nakazawa, M. Tanaka, and M. Fujita, Proc. Conf. Laser and Electro-Optics, CPD3 (2001).
29. K. Suzuki, H. Kubota, S. Kawanishi, M. Tanaka, and M. Fujita, Opt. Exp. **9**, 676 (2001).

30. T. Hasegawa, E. Sasaoka, M. Onishi, M. Nishimura, Y. Tsuji, and M. Koshiba, Opt. Exp. **9**, 681 (2001).
31. P. Yeh, A. Yariv, and E. Marom, J. Opt. Soc. Am. **68**, 1196 (1978).
32. N. J. Doran, and K. J. Blow, J. Lightwave Technol. **1**, 588 (1983).
33. S. G. Johnson, M. Ibanescu, M. Skorobogatiy, O. Weisberg, T. D. Engeness, M. Soljacic, S. A. Jacobs, J. D. Joannopoulos, and Y. Fink, Opt. Exp. **9**, 748 (2001).

(by T. Hasegawa)

Chapter 4
ANALYZING AND DESIGNING PHOTONIC CRYSTALS

4.1 INTRODUCTION

In this chapter, various methods for analyzing and designing photonic crystals are introduced. Since the possibility of controlling spontaneous emission using the PBG was first discussed in 1987, development of the methods for analyzing photonic crystals has progressed rapidly. In particular, the foundation for photonic crystal analysis has been strengthened by the work of theoretical researchers of solid state physics, including research groups individually headed by Leung, Zhang and Ho. They used the *plane wave expansion* (PWE) method to calculate energy bands, which is an optical version similar to the method used in solid state physics. In 1990, using this PWE method, Ho et al. calculated the band configuration of a diamond structure, and first showed that it is possible to obtain a complete PBG. In addition to the calculation of PBGs, the PWE can be applied to analyze the other aspects of photonic crystals, which have been discussed in Sections 2.4 and 2.5 as band engineering. The PWE calculation provides the contour map of frequency called the dispersion surface. It is the analogy to the Fermi surface of the electron system. In addition to providing a basic physical understanding of these phenomena, this technique is of general use in a wide range of application fields.

In the PBG and defect engineering in photonic crystals, which has been discussed in Section 2.2, artificial defects play an important role to manipulate photons. The photonic crystals with artificial defects are difficult to analyze within the limits of the PWE method, as they are incomplete crystals. Since the early stages of analysis, incomplete crystals have been analyzed using a supercell method, which introduces defects periodically. However, it takes a long time to calculate more complicated defect structures.

At around the same time, the *scattering matrix method* appeared on the scene. This method has an inherent advantage in its ability to derive the spatial distribution of electromagnetic fields with a high degree of accuracy in a short time, even in the case of large-scale nonperiodicity and defect introduction. However, the constraint on the structure of the photonic atom renders this method suitable for basic analysis only.

Since there are no limitations concerning the structures of photonic atom and defect, the *finite difference time domain (FDTD)* method is suitable for the analysis and design of the actual photonic crystal. Moreover, it is possible to analyze the time dependency of the optical pulse that propagates through a waveguide. As photonic bands can be calculated with a periodic boundary condition, the FDTD method is considered to be one of the principal methods for analyzing photonic crystals, along with the PWE method. Neither the finite element nor the beam propagation methods are suitable for analyzing photonic crystals in isolation. However, a CAD system, which combines these methods with the PWE method and/or the FDTD method, has been developed and is now commercially available as a software package.

(by H. Kosaka)

4.2 PLANE WAVE EXPANSION METHOD

4.2.1 Fundamental theory

The photonic band is a dispersion relation between frequency ω and wavevector (spatial frequency) k. The PWE method for photonic band calculation has made a substantial contribution to the development of photonic crystals. Further, the similarity between the Schrödinger equation for electrons and the wave equation for light has played an important role in this development. In band calculations for electrons, the periodic Coulomb potential in crystals is expressed by a multi-dimensional Fourier expansion. In other words, it is expressed as a plane wave potential pile-up. Suppose that an electron wave function is a Bloch function, this can also be expressed by a Fourier expansion. When substituted into the Schrödinger equation, the relation between energy E and kinetic momentum p is obtained. In free space, it is well known that it can be written as $E \propto p^2$.

The same relation can be given for lightwaves in a photonic crystal.[1-5] Figure 4.2.1 shows the real space of a photonic crystal with a period a against wavevector space. The Coulomb potential of an electron can be replaced by the dielectric constant potential $\varepsilon(r)$ of light and given by

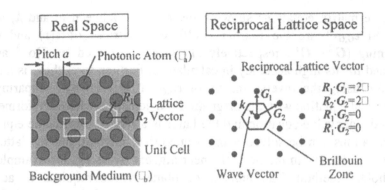

Figure 4.2.1 Photonic crystal structure and its corresponding reciprocal space concept.

$$\varepsilon(r) = \sum_{G} \varepsilon_G e^{jG \cdot r},\qquad(1)$$

where G is the reciprocal lattice vector in wavevector space, as introduced in solid state physics. The potential is given by the fundamental spatial frequency vector (fundamental reciprocal lattice vector) at Fourier expansion and its linear combination. If we use infinite numbers of G, a complete $\varepsilon(r)$ can be obtained. But in real calculations, a finite number (often called the plane wave number) should be used. Therefore, the degree of error in the calculation can be determined by this number. The wave function of an electric field or magnetic field can be given by

$$\Phi(r) = \sum_{G} \Phi_G e^{j(G+k) \cdot r},\qquad(2)$$

where the wavevector k is arbitrary, but if it is larger than $G/2$, k can be replaced by $k - G$, and the result will remain unchanged. Consequently, k values located outside of the Brillouin zone can all be replaced by k located inside the Brillouin zone, as shown in Fig. 4.2.1.

In a concrete band calculation, the two expressions mentioned above are substituted into the wave equation for light. However, depending on the choice of wave equation for either the electric field vector E or the magnetic field vector H, the equation to be solved will differ as [4]

$$\nabla \times [\nabla \times E(r,t)] = \omega^2 \varepsilon(r) E(r,t)$$
$$\rightarrow \quad (k+G) \cdot (k+G') \times E_{G'} = -\omega^2 \sum_{G'} \varepsilon_{G,G'} E_{G'},\qquad(3)$$

$$\nabla \times \left[\frac{1}{\varepsilon(r)} \nabla \times H(r,t) \right] = \omega^2 H(r,t)$$
$$\rightarrow \quad (k+G) \times \sum_{G'} \varepsilon_{G,G'}^{-1} (k+G') \times H_{G'} = -\omega^2 H_G.\qquad(4)$$

These are matrix equations showing the relation between ω and k, where $\varepsilon_{G,G'}$ and $\varepsilon_{G,G'}^{-1}$ are the Fourier coefficient matrices of $\varepsilon(r)$ and $\varepsilon^{-1}(r)$ concerning $(G - G')$, respectively. This was proposed by Ho [3] and is renowned for its high accuracy in calculation. Furthermore, there is another definition using the inverse matrix of $\varepsilon_{G,G'}$. It is seen by comparing the electric field equation with the magnetic field equation, where the former is a modified eigenvalue equation and the latter is a typical eigenvalue equation. Both equations can be solved using a numerical method, but the latter can readily be solved with the help of linear algebra by using, for example, the Householder method. The number of plane waves required to achieve adequate accuracy depends on the structural detail of the unit cell. In the case of a 2D photonic crystal with a simple circular optical atom, the accuracy of the band frequency from low frequency to the fourth band will be within 1% even for 40 plane waves. Similarly, in a 3D photonic crystal with a spherical atom, a plane wave number of 100 will give the same accuracy. When such accuracy is required for a high frequency range band, the number of plane waves will need to be increased depending on the required range. Moreover, when the atom structure is complicated, an increase in plane wave number will be required even for the low frequency range. In the case of a large scale and complicated element such as a supercell, which will be introduced in subsection 4.2.4, several thousands of plane waves are required.

4.2.2 Photonic band diagram

Figure 4.2.2(a) shows the band diagram of a 2D photonic crystal with triangular lattice free space. This free space means an infinite space with a periodic dielectric change infinitely approaching zero. In general, the photonic band diagram can be obtained by continuously connecting the obtained normalized frequency $\omega a/2\pi c$ $(= a/\lambda)$ by changing k between typical points in the Brillouin zone, where a is the lattice constant and c is the light velocity in vacuum. As the dielectric change is 0 in free space, there is a linear relation between ω and k. This point is quite different from electrons where it becomes a parabolic function. In contrast, the band diagram shown in Fig. 4.2.2(b) represents a case where there is a large change in dielectrics. There is a frequency range where band against wavevector in Γ-X, Γ-J or its intermediate direction does not exist. This is referred to as the stopband. The frequency range where the stopband superimposes in every direction is the 2D PBG. In this range, all light propagation and light emission is prohibited. The application of these properties to microlasers, resonators and waveguides is under development. The group velocity of light is given by $\partial\omega/\partial k$. That is, the slope of each band is proportional to the group velocity. In general, group velocity approaches zero at the band-edge. In addition, a number of low group

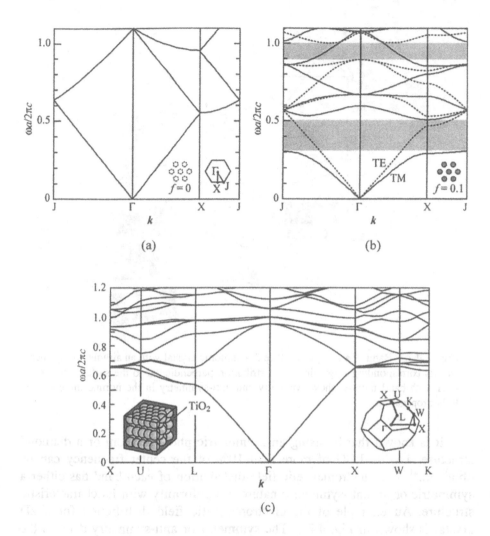

Figure 4.2.2 Examples of photonic bands. (a) 2D photonic crystal with triangular lattice arrangement. (b) 2D photonic crystal with 10% semiconductor cylinder content. (c) fcc lattice 3D photonic crystal made of TiO_2 with airholes in a close-packed arrangement.

velocities appear for 2D and 3D crystals, which are not observed in 1D crystals. This results in an enhanced interaction per unit length between the materials composing the photonic crystal and lightwaves. Therefore, properties such as optical gain, electro-optic and magneto-optic effects and nonlinear effects will be enhanced. For 3D crystals, the band can be drawn as it connects between typical points of a 3D Brillouin zone. The band for a fcc lattice is shown in Fig. 4.2.2(c). A PBG cannot be created for simple fcc structures.

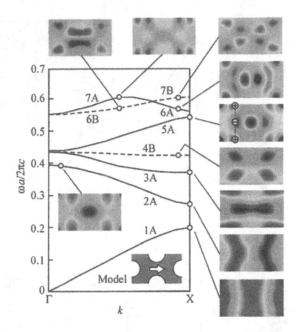

Figure 4.2.3 Band of a triangular lattice 2D photonic crystal with an airhole arrangement and its corresponding magnetic field distribution perpendicular to the surface. A and B after each band number show symmetry and anti-symmetry in the normal direction of light propagation.

It is known that by using an asymmetric photonic atom or a diamond structure, a wide PBG of more than 10% of the center frequency can be obtained.[3,6] The electromagnetic field distribution of each band has either a symmetric or an anti-symmetric nature in conformity with its characteristic structure. An example of the electromagnetic field distribution for a 2D crystal is shown in Fig. 4.2.3. The symmetry or anti-symmetry decides the coupling or uncoupling of the incident light from the exterior of the photonic crystal. This is important with regards to physics and its applications. In the case shown in Fig. 4.2.3, the fourth band has an anti-symmetric property towards incident light. Hence, an incident symmetric plane wave with this range of frequency will be totally reflected. It is known that the symmetry of each band can easily be predicted by the group theory.[7]

4.2.3 Dispersion surface

Calculation of $\omega a/2\pi c$ for all k in the Brillouin zone can be represented as contours. This is called the dispersion surface. An example of this contour for a 2D photonic crystal is shown in Fig. 4.2.4. The gradient at each point of the dispersion surface represents the Poynting vector (group velocity vector).

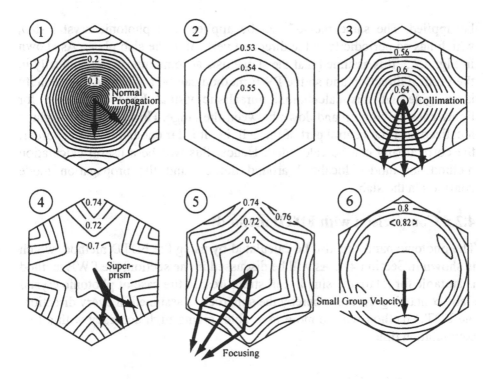

Figure 4.2.4 Triangular lattice 2D photonic crystal with an airhole arrangement and the dispersion surface of each band.

2D and 3D photonic crystals have complicated zone folding in comparison with 1D crystals. This means that, the dispersion surface plot will be distorted, an observation which has attracted much attention. The direction of the incident light into this range where the dispersion surface sharply changes will be altered significantly. This phenomenon is called a *superprism*, as described in Section 2.4. The application of this phenomenon as a wavelength filter is under investigation.[8] On the other hand, when light is injected into a range where the dispersion surface figure does not change, the direction of light is fixed. This phenomenon is called a *supercollimator*.[9]

4.2.4 Supercell calculation

In the previous calculations, infinite periodicity was assumed in all directions. However, this assumption cannot be used for photonic crystals, which have some defects introduced. In these cases, band calculation based on the supercell method is often applied. This concept is represented in Fig. 4.2.5. For crystals with a single defect, a set of several photonic crystal periods around the defect, redefined as an unit cell, is called a supercell. If it is assumed that such unit cells continue infinitely, then the PWE method can be applied. The same method can be applied to a photonic crystal slab, which has no periodicity in the direction vertical to the slab plane. As shown in Fig. 4.2.5, it is assumed that the slab exists at some interval periodically in the vertical direction and so the PWE method can be applied. As periodicity is imagined in this calculation, unrealistic virtual bands will exist for lightwaves moving up and down between the range. However, when light is localized near the central part of the cell, i.e. it is far enough from the border, the correct band can be calculated. Hence, this will be a useful calculation method for modes localized around defects and the propagation mode confined in the slab.

4.2.5 Comparison with FDTD method

The photonic band can also be calculated by using the FDTD method, which is shown in Section 4.4. As will be discussed in the section, the PWE method is suitable for relatively simple cell structures. In the case of photonic crystal slabs containing defects whereby the supercell is assumed in every direction, the FDTD method will be more favorable, giving high accuracy with a low computation time.

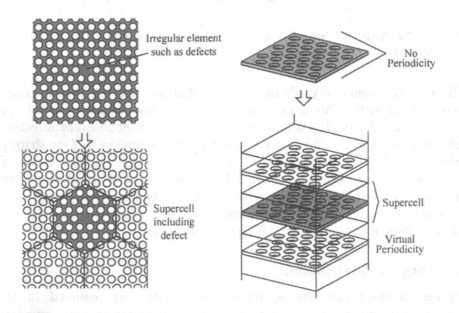

Figure 4.2.5 Concept of a supercell used for the PWE method. This is used for photonic crystals including irregular elements or those lacking periodicity in the specified direction.

References

1. K. M. Leung and Y. F. Liu, Phys. Rev. Lett. **65**, 2646 (1990).
2. Ze Zhang and S. Satpathy, Phys. Rev. Lett. **65**, 2650 (1990).
3. K. M. Ho, C. T. Chan and C. M. Soukoulis, Phys. Rev. Lett. **65**, 3152 (1990).
4. P. R. Villeneuve and M. Piché, Prog. Quantum Electron. **18**, 153 (1994).
5. J. D. Joannopoulos, R. D. Meade and J. N. Winn, Photonic Crystals, Princeton University Press, Princeton, 1995.
6. E. Yablonovitch, T. J. Gmitter and K. M. Leung, Phys. Rev. Lett. **67**, 2295 (1991).
7. K. Sakoda, Phys. Rev. B **52**, 7982 (1995).
8. H. Kosaka, T. Kawashima, A. Tomita, M. Notomi, T. Tamamura, T. Sato and S. Kawakami, Phys. Rev. B **58**, 10096 (1998).
9. H. Kosaka, T. Kawashima, A. Tomita, M. Notomi, T. Tamamura, T. Sato and S. Kawakami, Appl. Phys. Lett. **74**, 1212 (1999).

(by T. Baba)

4.3 SCATTERING MATRIX METHOD

4.3.1 What is the scattering matrix method?

Fundamentally, two ways to analyze photonic crystals exist. One is analytical and is carried out with the aid of calculators, for example, the PWE method. The other approach is numerical and is based on a differential method,[1] such as the FDTD method. The scattering matrix method belongs to the former methodology. In the scattering matrix method, the resource wave and the scattering objects are initially defined. The electromagnetic field of light at an arbitrary position is subsequently calculated from the summation of two waves: the wave arriving directly from the resource position and the wave that is scattered by an object. By means of Green's theorem and Graph's theorem of addition, the simple summation of these scattered waves can give the total scattered wave. Every scattered wave is represented in terms of cylindrical function expansion. The amplitude of each degree of cylindrical function, which is excited by waves from resource positions, is represented by the scattering matrix. It includes boundary conditions, i.e. the figure and refractive index of the objects (the name *scattering matrix method* is derived from this mechanism). The resource position, the scattering matrix, and the scattered wave are related to each other in simultaneous equations, and the excited amplitudes are solved using a calculator, immediately giving electromagnetic field distributions at arbitrary positions.

However, the following limitations apply: (i) This method can only be applied to 2D crystals. (ii) The objects must be isolated from each other, form a convex shape on the exterior and must be uniform on the interior.

Almost all calculation examples are restricted to circular objects because the calculation task is much simpler due to analytical scattering matrix solutions[2,3] induced for circular figures. (iii) Only static solutions of electromagnetic fields can be obtained, whilst the time response cannot be obtained. Moreover, nonlinear calculations are impossible under such an expansion representation.

4.3.2 Comparison with other methods

Despite restrictions such as those described above, the scattering matrix method is very useful for acquiring a precise static solution in a short time scale for a 2D crystal with a finite periodicity cycle under light incidence. The PWE method generally converges slowly, so better accuracy can be obtained using many plane waves. A complex structure requires over a thousand plane waves resulting in an enormous calculation process. Furthermore, it is generally difficult to treat an arbitrarily-sized object at a random position by the plane wave method. Although it is possible to treat an object with a random size at an arbitrary position using the FDTD method, this requires enormous time and memory resources for the calculation to obtain a high accuracy, since the accuracy depends on the size of the fine cubic cell dividing space. When the structure includes fine curves as is often the case in photonic crystals, the error cannot be suppressed below several % without a cell edge of under 1/80 of a wavelength in the medium. However, preparation of calculation resources is complex, therefore the calculation is finished using the conditions of 1/10 to 1/40 of a wavelength in many instances where the accuracy is not discussed in detail. On the other hand, the accuracy of the scattering matrix method depends on the number of cylindrical functions used in the calculation. The calculation converges rapidly for a round object and the error rate drops to under 3% with only −2 to +2 degree values.[4,5] The calculation volume is proportional to the square of the number of objects, so calculation for an object with a smaller crystal periodicity is more advantageous. But this is not such a serious problem, because in reality the practical characteristics of the photonic crystal are obtained by approximately five periodicities when the refractive index contrast between mediums is large. Since the calculation volume is not so affected by space size or cell size for mapping field distributions, a dense distribution is obtained covering a wider space than that produced using the FDTD method. For example, it takes under 30 seconds to calculate the distribution of an electromagnetic field around a crystal with a 5 × 10 periodicity using a 2 GHz Pentium IV personal computer.[5] This is shorter than 1/10 of the time required for the FDTD method using a Yee's cell edge with 1/20 of a wavelength. Nonuniform elements can be introduced easily as there is no limit in arrangement or size of the round objects.

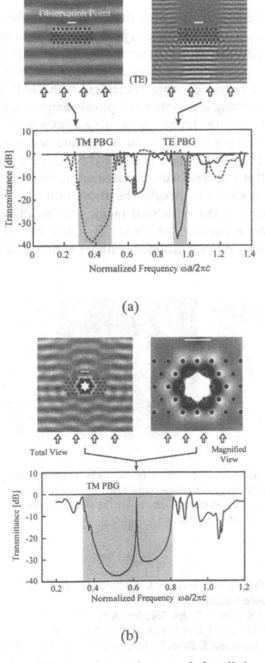

Figure 4.3.1 (a) Conformal 2D photonic crystal composed of a cylinder arrangement with 5 × 10 periodicity. (b) Electric field distributions of light around the crystal and Poynting power transmittance spectra where the crystal includes a single defect. For calculation, the refractive index of the semiconductors is assumed to be 3.5. The gray belt indicates a PBG induced by the PWE method.

4.3.3 Calculation examples

The electric field distribution of light and the transmission spectrum of a plane wave incident on a photonic crystal made of semiconductor cylinders arranged in a triangular lattice structure are shown in Fig. 4.3.1. Even in a crystal with five periodicities, a large decrease in transmittance of around 40 dB is found in a frequency range coincident with the PBG induced by the PWE method. A strong localization of optical energy and the generation of a resonant spectrum are found in a single defect. The electric field distributions of propagating light for a bending waveguide in a photonic crystal, a branching device and a directional coupler are shown in Fig. 4.3.2. It is predicted that complex photonic circuits can be formed in minute areas by utilizing these elements. Besides the above examples, a relation between the inhomogeneity of the cylindrical radius and the PBG effect have been investigated.[5] An unevenness of about 20 – 30% can be permitted, if the production target is set as a distorted cylinder radius, which halves the PBG effect.

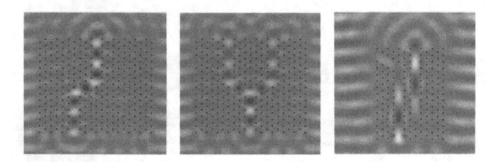

Figure 4.3.2 Electric field distributions of the light which is propagating in the various devices below. Left: Waveguides which contain two 60° bends. Center: a Y-branching device. Right: Directional coupler. Refractive indices are the same as those in Fig. 4.3.1. Polarization is TM. Wavelength is within the PBG.

References

1. D. Felbacq, G. Tayeb and D. Maystre: J. Opt. Soc. Am. A **11**, 2526 (1994).
2. G. O. Olaofe, Radio Science **5**, 1351 (1970).
3. H. A. Yousif and S. Köhler, J. Opt. Soc. Am. A **5**, 1085 (1988).
4. G. Tayeb and D. Maystre: J. Opt. Soc. Am. A **14**, 3323 (1997).
5. J. Yonekura, M. Ikeda and T. Baba, J. Lightwave Technol. **17**, 1500 (1999).

(by T. Baba)

4.4 FINITE DIFFERENCE TIME DOMAIN METHOD

4.4.1 General features

Finite difference time domain (FDTD) is a method that numerically calculates the space and time changes in the electromagnetic fields of light by using a finite difference method.[1] In this technique, an analytical finite space including the objective structure is divided into small cubic cells (small square cells for 2D case) called Yee's cells with edges of $\Delta x = \Delta y = \Delta z$, as shown in Fig. 4.4.1. Using the difference between the two principal Maxwell's equations, each x, y and z component (or x and y component for 2D crystals) of the electric field vector and the magnetic field vector of every cell is calculated in turn at every short time interval Δt. Of course, no change occurs in the field profile if an initial field does not exist. Therefore, some fields are excited at an appropriate position in space. The field propagating from that position finally reaches the periphery of the analytical space. Although the calculation is not executed in external space, this space must be treated as if it continues infinitely. A condition called the absorbing boundary is usually assigned to an electromagnetic field at the terminating boundary condition so as not to cause a reflection of the field and thereby return to the original analytical space. It is possible to set up an arbitrary figure in this model because the model is introduced by an assignment of the dielectric constant, which corresponds to the refractive index, and the conductivity, which corresponds to the absorption coefficient, of the medium material for each cell. The variation of the field in time can be obtained at an arbitrary position, allowing the frequency response to be readily induced by Fourier transformation of the variation. As will be described in the following section, this method can treat both the dispersion and the nonlinearity in

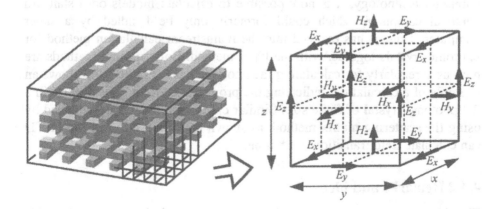

Figure 4.4.1 Calculation model and the Yee's cell.

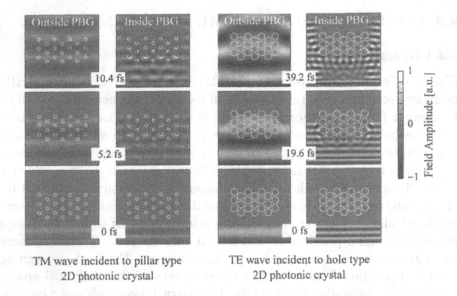

Figure 4.4.2 Calculation results of light propagation for two kinds of 2D photonic crystals.

object materials. With this multiuse capability, the FDTD method can be widely used in various application fields in both the research and development areas, including microwave apparatus as mobile-phones and optical devices. A number of FDTD software packages are also available.

Various calculation methods have been utilized for obtaining electromagnetic fields in photonic crystals. However, it is only recently that the FDTD method has been widely adopted to contribute to the development and implementation processes. This is because the method requires enormous computational resources, large amounts of memory and fast execution, as described below. However, due to the remarkable progress in computer technology, it is now possible to calculate models on a standard personal computer, which could formerly only be handled by a super computer. Thus, it has evolved into the mainstream calculation method for photonic crystals together with the PWE method. Both of these methods are now used regularly for calculating the photonic band. Figure 4.4.2 shows an example of a calculation, indicating the propagation of an optical pulse in a 2D photonic crystal of finite size. Similar calculation results are obtained by using the scattering matrix method, as shown in section 4.3. But the FDTD can calculate field transitions with time.

4.4.2 Remarks and ideas

The theory appears uncomplicated and the calculated electromagnetic fields seem quite reasonable using the FDTD method, since it only requires successive calculations, including the excitation and termination. Indeed, one should be careful about such aspect of the FTTD method. In order to perform a satisfactory calculation, consideration of accuracy is required. First, the calculation accuracy depends on the cell size. It is known that for a stable calculation, it is necessary to set the cell edge to below 1/10 of a wavelength in the medium material with the highest refractive index. For example, in a space with a refractive index of 3, a cell edge with a length below 50 nm is required to calculate a lightwave with a wavelength of 1.5 µm. In such an instance, a cell number of 200^3, that is, of 8 million is needed to analyze the total cubic space of a 10 µm edge. 56 megabytes of memory are required for allotting the six electromagnetic field components (three components in 2D case) and the dielectric constant of each cell in that space. A personal computer can handle such a scenario. However, when the structure in the analytical space includes a slope or a curved surface, a finer cell is required because the step approximation is executed by use of a cubic cell. Usually, a cell with an edge of 1/20 to 1/40 of the medium wavelength is used. An additional 8 to 64-fold increase in the amount of memory is required to improve the accuracy from two to fourfold. That means 448 MB to 3.6 GB. In reality, depending on the architecture of the program source though, over two times or more memory is often needed because of the absorbing boundary parameters, the parameters required in the middle of the calculation, and the definition of all values for higher precision. This level of computation is not appropriate for a personal computer. Moreover, the calculation time increases by 16 to 256 times (8 to 64 in 2D) because the unit time Δt used for analysis must be shortened in proportion to the cell definition when the cell is further divided into sections. A relatively good calculation technique is needed to successfully represent the flat areas in an electromagnetic field by coarse cells, using the rectangular cell, the sub-grid and the nonuniform mesh. However, the calculated results obtained by these methods are unstable at the point where different cell types are connected.

The performance of the absorbing boundary condition also affects the accuracy. There are two types of absorbing boundary conditions; one calculates fields so as to negate the reflection on the boundary surface, and the other suppresses the reflection by absorption of fields in a space around the analytical space. Mur's condition for the former and Berenger's perfect matched layer condition (PML) for the latter are prominent conditions. In reality, the reflection cannot be eliminated completely. Hence, several checkpoints are needed, such as a degree of reflection required in each calculation process, the optimum choice of parameters and a minimum amount of computer memory. Mur's condition is very attractive because it requires minimal calculation and memory consumption. But the degree of the reflection −suppression in power, −50 to −80 dB, is inferior to that in the

PML condition. Mur's condition is very useful for situations where such a suppression is considered to be sufficient. For example, this condition is worthwhile when observing the movement of the field profile over a relatively short time. On the other hand, the reflection suppression of PML is less than −150 dB in ideal conditions, which, with double precision, is under the calculation limit. Such a low level of reflection is very effective for resonant mode calculations and stable band calculations with periodic boundary, which will be introduced later. However, slow absorption of a field using a thick absorption layer is required to obtain sufficient suppression of reflection. In this case, the memory requirement sometimes becomes larger than the main body of the FDTD calculation. It is known that the performance of the reflection suppression degrades for all conditions when the boundary line of a complex structure comes into contact with the periphery of the analytical space. Discovering a way of including the continuity of the structure into the PML is still the theme of present research.

The appropriate excitation is important for obtaining the desired calculation result. This includes the position (or the areas) to be excited and the time function to be provided. Excitation at one point results in an equal excitation in the whole direction of the plane wave. Such an excitation method is effective in discovering new modes whose spatial distributions were not previously known, except for modes with zero amplitude at the exciting point (this mode is expressed as having a node at the excited point). In such cases, a technique with two or more exciting points, which have no symmetry in the structure, should be chosen. In the calculation of light propagation in a waveguide, it is inefficient to excite at just one point, because it involves a long path and calculation time until the excitation settles down to a stable waveguide mode. If a spatial excitation, which is very close to the waveguide mode, has been previously provided, the propagation length and the calculation time can be reduced because the mode will swiftly settle to the waveguide mode.

4.4.3 FDTD calculation using a periodic boundary

In addition to the capability of dealing with arbitrary situations using an absorption boundary covering the analytical space, the FDTD can also benefit from reduced calculation time by setting a small model with partial application of the special boundary condition. For example, it can be said that when a structure has a fourfold symmetry, the mode generated in that structure also has a fourfold symmetry. In the *x-y* plane, the calculated results in the second to fourth quadrants are expected to be symmetric (or anti-symmetric) to the results in the first quadrant. If a symmetric (or anti-symmetric) boundary condition is applied to the boundary line in contact with the second to fourth quadrants, the number of calculations is reduced to a quarter of the original number required. For a similar problem in 3D space, the number of calculations is reduced to 1/8 of the original quantity. In the case of a model structure supporting one periodicity with a boundary in the plane along the repeating direction of that model, a model with a virtually infinite scale can be calculated, bringing about a drastic reduction in the calculation time. In the most straightforward instance, light propagates parallel to the periodic boundary, and the simple assumption that the two field values on two lines of the periodic boundary are equal to each other can solve the problem. When the propagating direction of the lightwave is inclined to the periodic boundary, the phase shift should be applied to the field values on the same lines of the periodic boundary. In this case, all fields are calculated as complex numbers, and the phase shift $e^{-jk \cdot R}$ should be given where the wavevector in the oblique direction and the translation vector for one period are represented by k and R, respectively.

The technique of applying a phase shift can be used for the photonic band calculation in addition to the inclined light propagation calculation in the periodic structure. It is sometimes called the *order-N method* because of its fast convergence speed simply proportional to the Yee's cell size.[2] In the PWE method, after obtaining the frequency with respect to the wavevector k by solving the eigenvalue equations including structural information, the connection of ω obtained by the change in k generates the photonic band. In order to ensure adequate accuracy, many plane waves are necessary, and the calculation time increased in proportion to the square of the plane wave. More plane waves are required for more complex structures and this also results in the increased calculation time to achieve convergence. Calculating the photonic bands by using the FDTD method is an effective way to solve such a problem. An outline of the photonic band calculation is shown in Fig. 4.4.3. One small cell of the photonic crystal is chosen and the wavevector k is determined. It is subsequently surrounded with a periodic boundary, which provides the phase shift $e^{jk \cdot R}$. After providing a short pulse excitation at appropriate points, a vibration with a constant period is left by the lengthy continuous calculation. Fourier transformation gives rise to the eigen frequency ω with respect to k, and the photonic band is drawn by connecting these ω values with respect to various k. Figure 4.4.4 shows an example of a calculated band diagram in a line defect waveguide in a photonic crystal slab.[3] It is not easy to get a sufficiently converged solution for such a complex figure by the use of the PWE method. The incomplete modes, as indicated as gray areas in the figure, which are partially radiated to the upper and lower areas during propagation, can be made computable by using an absorbing boundary located at the upper and lower sections of the waveguide.

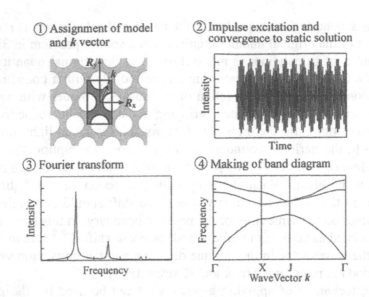

① Assignment of model and k vector

② Impulse excitation and convergence to static solution

③ Fourier transform

④ Making of band diagram

Figure 4.4.3 Outline of the photonic band calculation using the FDTD method.

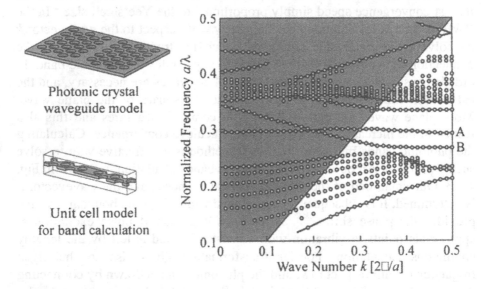

Photonic crystal waveguide model

Unit cell model for band calculation

Figure 4.4.4 Waveguide model with a line defect introduced in the photonic crystal slab and the calculated photonic band. Open circles indicate the results for waveguides assuming that the periodical boundaries are at the front and rear and at the right and left surfaces along with the light propagating direction. Solid lines indicate the results with the periodical boundaries just at the front and rear surfaces. Other areas are the PML absorption boundaries. Lines which are indicated by A and B represent the band structure of the waveguide. The gray zone corresponds to the lightcone representing light leakage through the upper and lower areas.

4.4.4 Special FDTD calculations

The behavior of an electromagnetic field can be calculated by using the orthogonal function expansion found in many numerical techniques. That is to say, linearity is the major premise for most numeric techniques. On the other hand, the FDTD method can be used for nonlinear problems because it does not contain any linear expansion representation. At present, calculation techniques mainly applicable to the third order nonlinearity are known.[4] By Maxwell's equation, the time difference of a displacement vector D, given by $D = \varepsilon E$, is induced from the rotation of the magnetic field H, where E is the electric field and ε is the dielectric constant. The value of E is deduced directly from the value of H when ε is fixed. It should be noted that when there is a third order nonlinear effect, which is expressed by an equation such as $D = \varepsilon_0(n^2 + 4\chi^{(3)}|E|^2)E$ in SI units, E is not so easily derived. Here, E is solved by going through ordered steps that first require deduction of D from H and then can be solved by the cubic equation obtained by the above relation between D and E. The generation of a third (or odd order) higher frequency harmonic wave must be considered in the analysis of these cubic equations. The calculation becomes unstable, if we do not set appropriately fine Yee's cells, say at least 1/3 of the usual length, because light with a wavelength of 1/3 of the fundamental wavelength is generated. Although there are not many calculated examples, it has been shown that a nonlinear response, which exhibits a reduction of transmittance and a steep saturation curve, is generated due to an expansion of nonlinearity occurring at the band-edge in 2D photonic crystals.[5]

In the usual FDTD method, ε is expected to be constant with respect to the frequency. In practice, materials always exhibit some dispersion. In particular, a reasonable solution cannot be obtained without introducing dispersion for materials with intensively dispersive characteristics such as metals. The relation between D and E is represented as $D(\omega) = \varepsilon(\omega)E(\omega)$. The inverse Fourier transformation of this equation can be represented as the convolution integral form $D(t) = \int \varepsilon(t')E(t - t')\,dt'$. All previous data on $E(t)$ can generate E at the next time step, since if $\varepsilon(\omega)$ of the material is known in advance, then $\varepsilon(t)$ is always obtained by the inverse Fourier transform of $\varepsilon(\omega)$. In reality, all data obtained previously is not necessarily required for the actual calculation, and the new E can be calculated by an inductive method using data from time field data at $t - \Delta t$ and $t - 2\Delta t$. For metals, the calculation, including the well known second Lorentzian dispersion, is actually executed.

References

1. Although there are numerous text-books available, the following book covers almost all of the techniques: A. Taflove, S. C. Hagness (Ed.), Computational Electrodynamics: The Finite-Difference Time-Domain Method — 2nd Ed., Artech House Publishers, 2000.
2. C. T. Chan, Q. L. Yu and K. M. Ho, Phys. Rev. B **51**, 16635 (1995).
3. T. Baba, A. Motegi, T. Iwai, N. Fukaya, Y. Watanabe and A. Sakai, IEEE J. Quantum Electron. **38**, 743 (2002). There are many calculated examples like this available.
4. P. Tran: Opt. Lett. 21 (1996) 1138.
5. T. Baba and T. Iwai, Jpn. J. Appl. Phys. **42**, (2003).

(by T. Baba)

4.5 FINITE ELEMENT METHOD

4.5.1 Analyses of photonic crystal fibers

In addition to the rapid progress made in the practical use of photonic crystal fiber (PCF) research, the development of an analytical method to evaluate the transmission characteristics has advanced intensively. The equivalent refractive index method[1] is one of the simplest methods, and is able to express theoretically almost all the interesting characteristics of a PCF. However, it is difficult to evaluate precisely the transmission characteristics of a PCF including the chromatic dispersion and modal birefringence, which strongly depends on the size, shape, number and arrangement of holes. Therefore, in addition to methods that are function expansion based, for example, the PWE method,[2] Hermite-Gauss function expansion method[3] and multipole method,[4] the introduction of numerical approaches such as the finite element method (FEM),[5-8] the FDTD method[9-12] and the beam propagation method,[13] which can handle full vector analysis, has progressed very quickly. They are all in conformity with full vector analyses.

Moreover, PCFs can be divided into holey fibers and PBG fibers according to the waveguiding principle.[14,15] Holey fibers, which have holes in the cladding region, utilize the difference of refractive indices between the core and the cladding so as to confine the light by total reflection. In contrast, PBG fibers with PBG in the cladding confines the light in the core part by means of a Bragg reflection effect. Known as an extremely general method, the finite element method is able to deal freely with various PCF analyses.[16] In this section, the finite element method[17] for full vector wave analysis (VFEM) is introduced, as well as the evaluated results of the modal birefringence and the chromatic dispersion characteristics of holey fibers.

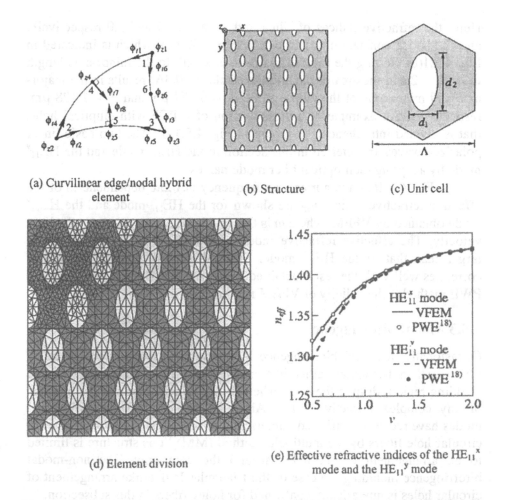

(a) Curvilinear edge/nodal hybrid element

(b) Structure

(c) Unit cell

(d) Element division

(e) Effective refractive indices of the $HE_{11}{}^x$ mode and the $HE_{11}{}^y$ mode

Figure 4.5.1 Curvilinear elements and a photonic crystal fiber comprised of elliptical airholes.

4.5.2 Finite element method

Although a triangular element is generally used for 2D FEM analysis, a curvilinear element which corresponds closely to the hole patterns of various shapes such as circles and ellipses, is introduced in this subsection. The curvilinear edge/nodal hybrid element[17] used in VFEM analysis is shown in Fig. 4.5.1(a). The transverse electric or magnetic fields in the PCF cross-section are approximated in every element using eight edge variables, $\phi_{t1} - \phi_{t8}$, and the longitudinal electric or magnetic fields in the propagating direction are approximated using six nodal variables, $\phi_{z1} - \phi_{z6}$, again in every element.

As is shown in Fig. 4.5.1(b), a PCF with elliptical holes arranged as a triangular lattice in silica glass is used to confirm the validity of VFEM.

Here, the refractive indices of silica and air are 1.45 and 1.0 respectively, and the lattice constant of a unit cell is $\Lambda = 2.58$ μm, which is indicated in Fig. 4.5.1(c). Setting the ratio of major-axis length to the minor-axis length to $d_2/d_1 = 2$ and the area of the elliptical holes to $0.3\Lambda^2$ results in the major-axis and minor-axis of the ellipse to be $d_1 \approx 0.564$ μm and $d_2 \approx 1.128$ μm, respectively. An example of a cross-section of a PCF with elliptical holes that is divided into elements is shown in Fig. 4.5.1(d). The two basic linear polarized waves are referred in this section to the $HE_{11}{}^x$ mode and the $HE_{11}{}^y$ mode by adopting their optical fiber mode names.

In Fig. 4.5.1(e), the normalized frequency $\omega\Lambda/2\pi c$ dependencies of the effective refractive indices n_{eff} are shown for the $HE_{11}{}^x$ mode and the $HE_{11}{}^y$ mode obtained by VFEM, where ω is the angular frequency and c is the light velocity. The effective refractive index of the $HE_{11}{}^x$ mode appears to be larger than that of the $HE_{11}{}^y$ mode. Since the result obtained by VFEM correlates well with the result obtained from vector wave analysis using the PWE method,[18] the validity of VFEM is confirmed.

4.5.3 Modal birefringence

Occasionally, a modal birefringence is observed in experiments in holey fibers with a triangular lattice-like arrangement of circular holes in the cladding region, despite the established theory which stipulates the absence of any complex refractive index. Although two fundamental orthogonal modes have recently confirmed numerically to degenerate from each other in circular hole fibers by the multipole method (MM),[4] this structure is limited to an arrangement of six holes around the core only. The non-modal birefringence including the case of the triangular lattice-like arrangement of circular holes is numerically confirmed for holey fibers in this subsection.

A holey fiber with a structure, as shown in Fig. 4.5.2(a) is considered first. The six holes are symmetrically arranged around the core. The effective refractive indices $n_{eff}{}^x$ and $n_{eff}{}^y$, respectively, for the $HE_{11}{}^x$ and $HE_{11}{}^y$ modes and the modal birefringence $n_{eff}{}^x - n_{eff}{}^y$ for the HF are shown in Fig. 4.5.2(b). Here, $\Lambda = 6.75$ μm, $d = 5$ μm, and $\lambda = 1.55$ μm are all defined. From the symmetry of the structure, the analytical region is determined as a quarter of the fiber cross-section and the size of that region R is 40.5 μm. A modal birefringence at the degree of freedom 99687 is estimated in the order of 1×10^{-8} which is similar to the results obtained from MM.[4]

Next, a holey fiber with a structure, as shown in Fig. 4.5.3(a), is evaluated. The circular holes are arranged in a triangular lattice in the cladding region. The result for this fiber is shown in Fig. 4.5.3(b) when assuming $\Lambda = 2.26$ μm, $d = 1.51$ μm, and $\lambda = 0.8$ μm.[9] The analytical region is defined as $X = 10.17$ μm and $Y = 9.13$ μm. As the modal birefringence at

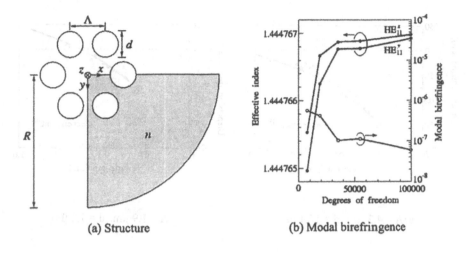

(a) Structure   (b) Modal birefringence

Figure 4.5.2 Holey fiber with a one-ring airhole array.

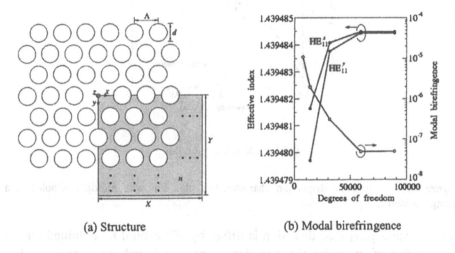

(a) Structure   (b) Modal birefringence

Figure 4.5.3 Holey fiber with multiple airholes in a triangular lattice.

the degree of freedom 88345 is estimated as approximately 4.9×10^{-8}, it is concluded that there is no modal birefringence, even after numerical error is taken into consideration.

The holey fiber in Fig. 4.5.2 and 4.5.3 has a six-fold rotational symmetry, hence there are 12 linearly polarized degenerated modes whose polarization directions are different from each other by $60°$, i.e. six for the HE_{11}^{x} modes and six for the HE_{11}^{y} modes. These degenerate modes are expressed as a combination of the HE_{11}^{x} and HE_{11}^{y} modes. If there is a modal birefringence between the orthogonal polarized modes, the linearly polarized

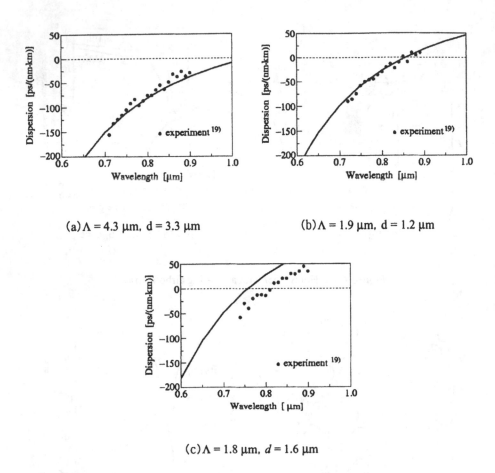

(a) $\Lambda = 4.3$ μm, $d = 3.3$ μm   (b) $\Lambda = 1.9$ μm, $d = 1.2$ μm

(c) $\Lambda = 1.8$ μm, $d = 1.6$ μm

Figure 4.5.4 Chromatic dispersion characteristics of a HF with multiple airholes in a triangular lattice.

waves whose polarized direction is offset by 60° cannot be obtained due to the rotation of the polarized direction. There exist such degenerate modes, thus we conclude that the modal birefringence does not exist in a holey fiber with six-fold rotational symmetry. The modal birefringence, which is sometimes observed in experiments, is therefore thought to be attributed to the incompleteness of the structure itself.

4.5.4 Chromatic dispersion characteristics

Figure 4.5.4 shows a calculated example of the chromatic dispersion characteristics with respect to the holey fiber in Fig. 4.5.3. The chromatic dispersion σ [ps/km·nm] is acquired as the sum of the material dispersion σ_m and the waveguide dispersion σ_w. The waveguide dispersion is derived using

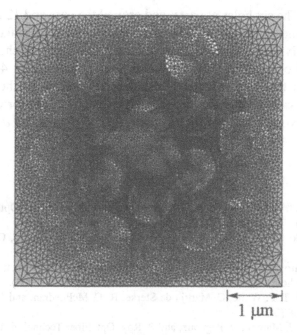

(a) Finite element mesh

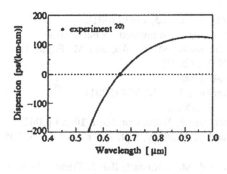

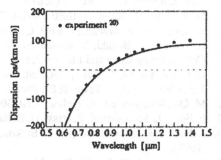

(b) Chromatic dispersion (real model simulation)

(c) Chromatic dispersion characteristics (Λ = 1.85 μm and d = 1.1 μm).

Figure 4.5.5 Holey fiber with a two-ring airhole array. (a) Finite element mesh. (b) Chromatic dispersion (real model simulation). (c) Chromatic dispersion characteristics (Λ = 1.85 μm and d = 1.1 μm).

the effective refractive index obtained from the guided mode analysis and the material dispersion is deduced using Sellmeir's dispersion equation for pure silica. In Fig. 4.5.4, Λ = 4.3 μm and d = 3.3 μm for (a), Λ = 1.9 μm, d = 1.2 μm for (b), and Λ= 1.8 μm, d = 1.6 μm for (c) are assumed. The calculated results agree well with the experimental results with the exception of the case (c).[19] Next, the holey fiber structure shown in Fig. 4.5.5(a) is considered. The six holes are arranged around the core and 12 more holes are arranged in the outer region. A finite element mesh for simulating the actual holey fiber is also indicated. The chromatic dispersion characteristics of structure (a) obtained by the simulation are shown in Fig. 4.5.5(b), which demonstrates that this simulated result agrees well with the experimental result.[20] The calculated result and the experimental result for the holey fiber possessing two rings of holes arrays, when $\Lambda = 1.9$ μm and $d = 1.2$ μm, are shown in Fig. 4.5.5(c).

References

1. J. C. Knight, T. A. Birks, P. St. J. Russell, and J. P. de Sandro, J. Opt. Soc. Am. A **15**, 748 (1998).
2. A. Ferrando, E. Silvester, I. J. Miret, P. Andres, and M. V. Andres, Opt. Lett. **25**, 276 (1999).
3. T. M. Monro, D. J. Richardson, N. G. R. Broderick, and P. J. Bennett, "Modeling large air fraction holey optical fibers", J. Lightwave Technol. **18**, 50 (2000).
4. M. J. Steel, T. P. White, C. Martijn de Sterke, R. C. McPhedran, and L. C. Botten, Opt. Lett. **26**, 488 (2001).
5. F. Brechet, J. Marcou, D. Pagnoux, and P. Roy, Opt. Fiber Technol. **6**, 181 (2000).
6. T. Hasegawa, E. Sasaoka, M. Onishi, M. Nishimura, Y. Tsuji, and M. Koshiba, Proc. Opt. Fiber Commun. Conf., PD5-1 (2001).
7. M. Koshiba and K.Saitoh, IEEE Photon. Technol. Lett. **13**, 1313 (2001).
8. K. Saitoh and M. Koshiba, IEEE J. Quantum Electron., submitted for publication.
9. H. Kubota, K. Suzuki, S. Kawanishi, M. Nakazawa, M. Tanaka, and M. Fujita, Proc. Conference on Laser and Electro-Optics, CPD3-1 (2001).
10. G. E. Town and J. T. Lizier, Opt. Lett. **26**, 1042 (2001).
11. J. T. Lizier and G. E. Town, IEEE Photon. Technol. Lett. **13**, 794 (2001).
12. M. Qiu, Microwave Opt. Technol. Lett. **30**, 327 (2001).
13. F. Fogli, L. Saccomandi, P. Bassi, G. Bellanca, and S. Trillo, Opt. Expr. **10**, 54 (2002).
14. J. Broeng, D. Mogilevstev, S. E. Barkou, and A. Bjarklev, Opt. Fiber Technol. **5**, 305 (1999).
15. T. A. Birks, J. C. Knight, B. J. Mangan, and P. St. J. Russell, IEICE Trans. Electron. **E84-C**, 585 (2001).
16. M. Koshiba, IEICE Trans. Electron., accepted for publication.
17. M. Koshiba and Y. Tsuji, J. Lightwave Technol. **18**, 737 (2000).
18. M. J. Steel and R. M. Osgood, Jr., Opt. Lett. **26**, 229 (2001).
19. D. Ouzounov, D. Homoelle, W. Zipfel, W. W. Webb, A. L. Gaeta, J. A. West, J. C. Fajardo, and K. W. Koch, Opt. Commun. **192**, 219 (2001).
20. J. C. Knight, J. Arriaga, T. A. Birks, A. Ortigosa-Blanch, W. J. Wadsworth, and P. St. J. Russell, IEEE Photon. Technol. Lett. **12**, 807 (2000).

(by M. Koshiba)

Chapter 5

EXAMPLES OF VARIOUS PHOTONIC CRYSTAL APPLICATIONS

5.1 INTRODUCTION

In this chapter, proposals for and examples of photonic crystal applications are reviewed, referring to the relevant publications from the body of literature covering this field of investigation. This chapter is composed of the following 11 sections:

Lasers
Light-emitting diodes (LEDs)
Resonators and filters
Waveguides
Fibers
Prisms and polarizers
Photonic integrated circuits
Nonlinear devices
Tunable crystals and optical switches
Antennas and electromagnetic wave techniques
Miscellaneous

Although many other papers treat 1D crystals such as diffraction gratings and multilayer films as photonic crystals, we concentrate mainly on 2D and 3D photonic crystals in this chapter. Appropriate articles from each relevant research organization are introduced in accordance with the application. Note that all the published papers are not covered. Actually, this session was mainly summarized by graduate students in Kyoto Univ. and Yokohama National Univ. based on their interests.

Note that the copyright of each figure in this chapter is taken from the corresponding author and publisher of each paper indicated directly by the figure number.

5.2 LASERS

Research here is divided into two types of investigation according to the method of laser construction and application performance. Subsection 5.2.1 discusses the utilization of a defect level in the PBG and Subsection 5.2.2 discusses lasers utilizing a band-edge with a zero group velocity.

5.2.1 Lasers utilizing defect levels in the PBG

The introduction of defects into photonic crystals results in the creation of a PBG in which a defect level is generated, and consequently light is confined to the localized defect state. Clear demonstration for such a defect-mode lasing was at first demonstrated in 1999, although thresholdless lasing was still not realized.

(1) "Donor and acceptor modes in photonic band structure"

E. Yablonovitch, T. J. Gmitter, R. D. Meade, A. M. Rappe, K. D. Brommer, and J. D. Joannopoulos (Bell Commun. Res., etc., USA), Phys. Rev. Lett. **67**, 3380 (1991).

A 3D photonic crystal of Yablonovite structure was fabricated from dielectrics using a mechanical machining process in three directions. A cavity was constructed from this structure by introducing defects (adding or removing periodically arranged elements).

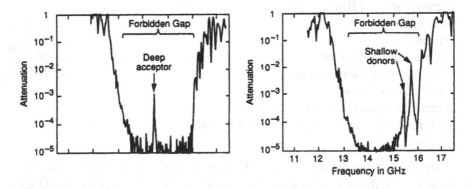

Figure 5.2.1(1) The observed defect levels in a Yablonovite crystal, with defect introduced into a structure with millimeter scale order periodicity.

(2) "Photonic band-gap structures"

E. Yablonovitch (UCLA, USA), J. Opt. Soc. Am. B **10**, 283 (1993).

A single mode emission diode is proposed by utilizing a PBG crystal as a cavity into which a defect level is introduced. The idea describes a new device which possesses both an LED advantage (thresholdless, high reliability) and a semiconductor laser advantage (coherence, high efficiency).

Single-Mode Light-Emitting Diode

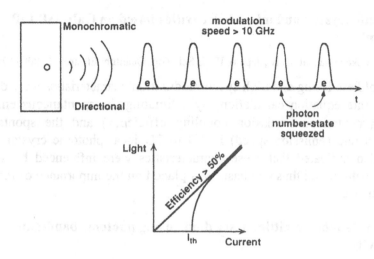

Figure 5.2.1(2) Characteristics of a photonic crystal light-emitting device.

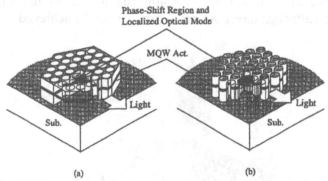

Figure 5.2.1(3) Composition of the suggested laser device, made of a circular hole and circular column 2D photonic crystal.

(3) "Fabrication and photoluminescence studies of GaInAsP/InP 2-dimensional photonic crystals"

T. Baba, and T. Matsuzaki (Yokohama Nat. Univ., Japan), Jpn. J. Appl. Phys. **35**, 1348 (1996).

EB lithography and RIBE etching methods using a methane gas system were carried out on a compressively-strained quantum well wafer of GaInAsP/InP, in order to fabricate a 2D photonic crystal composed of a pillar arrangement. An increase and decrease of light emission for lateral direction from this structure was observed. Moreover, they propose a laser device constructed in this way, using a 2D photonic crystal structure.

"Photonic crystals and microdisk cavities based on GaInAsP/InP system"

T. Baba (Yokohama Nat. Univ., Japan), IEEE Sel. Top. Quantum Electron. **3**, 808 (1997).

Thresholdless lasing and high speed modulation characteristics were derived from a rate equation calculation, by estimating the spontaneous emission factor (spontaneous emission coupling efficiency) and the spontaneous emission rate (emission speed) in 1D to 3D ideal photonic crystal lasers. They also indicated that these characteristics were influenced by surface recombination, and thus emphasis was placed on the importance of reducing the influence.

(4) "Novel surface emitting laser diode using photonic band-gap crystal cavity"

H. Hirayama, T. Hamano, and Y. Aoyagi (Riken, Japan), Appl. Phys. Lett. **69**, 791 (1996).

A surface-emitting laser using a sheet type defect introduced into a 3D photonic crystal is proposed. It is theoretically shown that high output power can be expected from a wide emission region due to the sheet defect configuration although thresholdless current cannot be achieved.

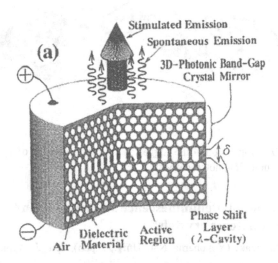

Figure 5.2.1(4) Surface-emitting laser by using sheet type defect in a 3D photonic crystal.

(5) "Lasers incorporating 2D photonic bandgap mirrors"

J. O'Brien, O. Painter, R. Lee, C. C. Cheng, A. Yariv, and A. Scherer (Caltech, USA), Electron. Lett. **32**, 2243 (1996).

A 2D photonic crystal was fabricated as mirrors by the dry-etching of a GaAs wafer. Lasing oscillation was observed in a device with the crystals as both mirrors instead of the cleaved facets. The device can be considered a 2D version of a DRB laser, with a deep diffraction grating produced by dry-etching as was previously reported.

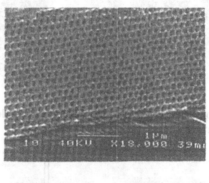

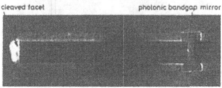

Figure 5.2.1(5) Fabricated circular hole 2D photonic crystal, and a stripe laser using this crystal as an edge mirror.

(6) "Photonic bandgap disk laser"

R. K. Lee, O. J. Painter, B. Kitzke, A. Scherer, and A. Yariv (Caltech, USA), Electron. Lett. **35**, 569 (1999).

Lasing oscillation of photonic crystal formed in a thin film slab on semiconductor was observed when the center part without crystal was photopumped. This is also considered as DFB type laser. The dimension of the emission region was 10 – 15 µm at this stage.

(7) "Room temperature photonic crystal defect lasers at near- infrared wavelengths in InGaAsP"

O. J. Painter, A. Husain, A. Scherer, J. D. O'Brien, I. Kim, and P. D. Dapkus (Caltech, etc. USA), J. Lightwave Technol. **17**, 2082 (1999).

The dimensions of the light emitting region were designed to be as small as a single defect in the above-mentioned laser (see above (6)). Using this device, pulsed lasing oscillation was realized by photopumping at room temperature. Light emission in a short wavelength range, which is shown in the figure, is due to the band-edge. The data agree well with numerical calculation results by the FDTD method.

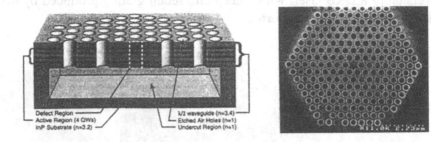

Figure 5.2.1(7-1) Conceptual figure of a 2D photonic crystal defect mode laser and a photograph of its top side.

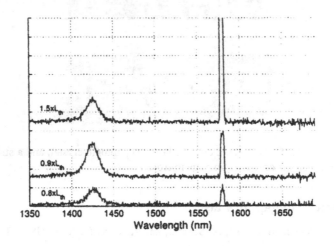

Figure 5.2.1(7-2) Lasing characteristics.

(8) "Effect of a photonic band gap on the threshold and output power of solid-state lasers and light-emitting diodes"

M. H. Szymanska, A. F. Hughes, and E. R. Pike (King's College London, UK), Phys. Rev. Lett. **83**, 69 (1999).

Both the threshold current and output power were compared theoretically for the cases of a conventional laser and a photonic crystal laser with bandgap effect. The threshold was found to be lower, and the power was seen to be increased in the near threshold region for the photonic crystal laser. However, the difference in power for the above threshold regime is found to be small.

(9) "Room temperature triangular-lattice two-dimensional photonic band gap lasers operating at 1.54 μm"

J. K. Hwang, H. Y. Ryu, D. S. Song, I. Y. Han, H. W. Song, H. K. Park, and Y. H. Lee (KAIST, etc., Korea), Appl. Phys. Lett. **76**, 2982 (2000).

A 2D photonic crystal of circular holes was formed on a fused substrate which is composed of GaInAsP/InP film on alumina. Light was confined within defects by total reflection for the vertical direction, and by photonic crystal mirrors for the horizontal direction.

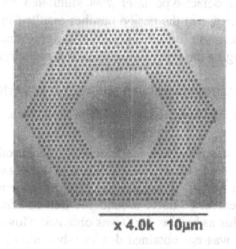

x 4.0k 10µm

Figure 5.2.1(9-1) SEM image from the upper side of the 2D photonic crystal laser

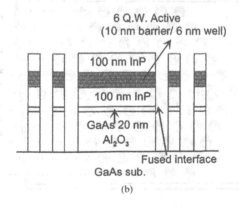

Figure 5.2.1(9-2) Side view of the device

"Continuous room-temperature operation of optically pumped two-dimensional photonic crystal lasers at 1.6 μm"

Same group, IEEE Photon. Technol. Lett. **12**, 1295 (2000).

This is a subsequent report to the previous one. Continuous wave lasing at room temperature was achieved using a hexagonal cavity with a diameter of ~10 μm.

(10) "Threshold gain and single-mode oscillation of two-dimensional photonic bandgap defect lasers"

N. Susa (NTT, Japna), IEEE J. Quantum Electron. **37**, 1420 (2001).

The behavior of a defect-type laser was simulated by changing various parameters. An increase in the period number resulted in an increase of the Q-value and a decrease of the threshold gain. Calculations were also performed by changing the size of the defect and/or the refractive index.

(11) "High quality two-dimensional photonic crystal slab cavities"

T. Yoshie, J. Vuckovic, A. Scherer, H. Chen, and D. Deppe (Caltech, etc., USA), Appl. Phys. Lett. **79**, 4289 (2001).

A donor mode nano-cavity was fabricated by the introduction of a single defect in a 2D photonic crystal slab. A quantum dot possessing an emission wavelength of 1.1 μm to 1.3 μm was used as an emission source. By changing the placement of holes around the defects (fractional dislocation geometry), a Q-value as high as 2800 was obtained. However, the theoretical Q-value of 30000 was not obtained due to fabrication processing accuracy limits, and problems in optimization.

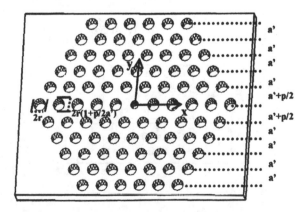

Figure 5.2.1(11-1) Typical structure of a 2D photonic crystal

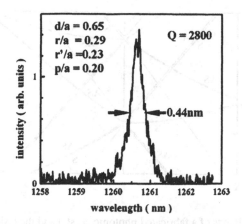

Figure 5.2.1(11-2) Photoluminescence spectrum of the sample

(12) "Analysis and design of single-defect cavities in a three-dimensional photonic crystal"

M. Okano, A. Chutinan, and S. Noda (Kyoto Univ., Japan), Phys. Rev. B **66**, 165211 (2002).

Theoretical investigation of the properties of a single-defect cavity created by adding dielectric material to a 3D photonic crystal was performed by utilizing the plane wave expansion method and the 3D FDTD method. Design rules for developing a single mode high-Q cavity in a 3D photonic crystal are supplied.

5.2.2 Lasers utilizing the band-edge

The details on this mechanism is described in Section 2.2, where phenomena that the group velocity becomes zero at the band-edges are utilized. Note that some of the reports are based on the different mechanisms.

(1) "A two-dimensional photonic crystal laser"

K. Inoue, M. Sasada, J. Kawamata, K. Sakoda, and W. H. Haus (Hokkaido Univ., Japan), Jpn. J. Appl. Phys. **38**, L157 (1999).

This paper describes that the laser operation of a 2D photonic crystal comprised of a circular hole arrangement. Narrowing of the spectrum was observed by photopumping the 2D crystal with a dye in the circular holes of a fiber plate. The lasing mechanism may be different from the band-edge effect according to the authors.

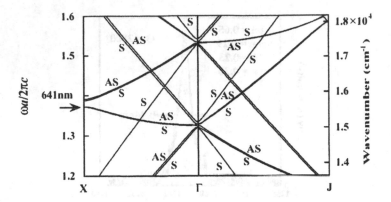

Figure 5.2.2(1) Bands of a fabricated photonic crystal and the lasing wavelength.

(2) "Coherent two-dimensional lasing action in surface-emitting laser with triangular-lattice photonic crystal structure"

M. Imada, S. Noda, A. Chutinan, T. Tokuda, M. Murata, and G. Sasaki (Kyoto Univ., Japan), Appl. Phys. Lett. **75**, 316 (1999).

A band-edge laser diode of a triangular-lattice 2D photonic crystal was successfully achieved by current injection. The utilization of a higher frequency Γ point allowed the surface emitting type laser action. Large area coherent lasing was confirmed in a narrow output beam.

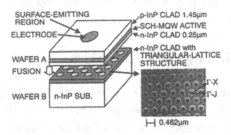

Figure 5.2.2(2-1) Schematic structure of surface-emitting 2D photonic crystal laser.

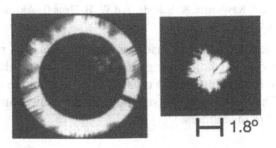

Figure 5.2.2(2-2) Observed near-field pattern (left) and far-field pattern (right).

(3) "Laser action from two-dimensional distributed feedback in photonic crystals"

M. Meier, A. Mekis, A. Dodabalapur, A. A. Timko, R. E. Slusher, J. D. Joannopoulos, and O. Nalamasu (Lucent Bell Labs., etc., USA), Appl. Phys. Lett. **74**, 7 (1999).

The fabrication of a laser using a 2D triangular lattice photonic crystal was described, along with an estimation of its lasing characteristics. The lasing oscillation was obtained by photopumping of an organic film containing a dye, which was formed on a glass circular hole arrangement.

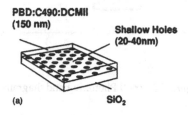

Figure 5.2.2(3-1) Device structure.

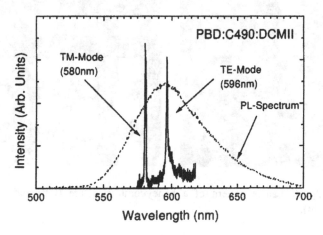

Figure 5.2.2(3-2) Oscillation spectrum.

"Emission characteristics of two-dimensional organic photonic crystal lasers fabricated by replica molding"

Same group (Lucent Bell Labs., USA), J. Appl. Phys. **86**, 3502 (1999).

This is a report subsequent to the above one. It describes a laser structure using a 2D photonic crystal of either a honeycomb or a square arrangement, fabricated by a replica molding technique. Lasing occurred by photopumping, and a far-field image with good symmetry was confirmed.

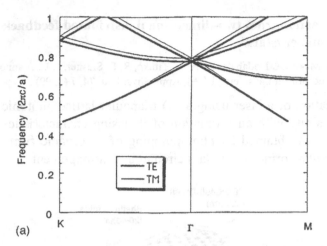

(a)

Figure 5.2.2(3-3) Photonic band diagram.

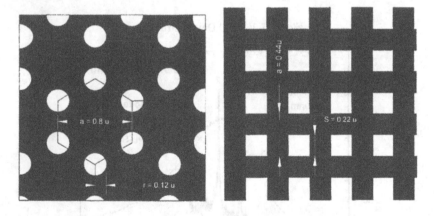

Figure 5.2.2(3-4) Lattice pattern of a honeycomb structure (left) and a square structure (right).

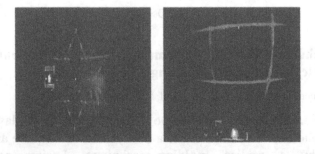

Figure 5.2.2(3-5) Far-field images of the lasers. Triangular lattice laser (left) and square lattice laser (right).

(4) "Polarization mode control of two-dimensional photonic crystal laser by unit cell structure design"

S. Noda, M. Yokoyama, M. Imada, A. Chutinan, and M. Mochizuki (Kyoto Univ., Japan), Science **293**, 1123 (2001).

Polarization mode selection in a 2D photonic crystal laser based on band-edge engineering is demonstrated by controlling the geometry of the unit cell structure. It is shown that unprecedented type of lasers with perfect single modes from the viewpoints of longitudinal, lateral, and polarization.

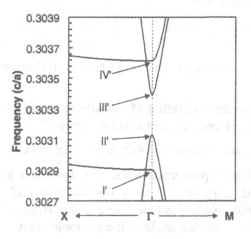

Figure 5.2.2(4-1) Bands of square lattice crystal with elliptical photonic atoms.

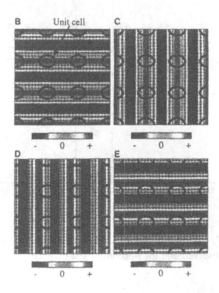

Figure 5.2.2(4-2) Calculated electromagnetic field distribution at individual band-edges.

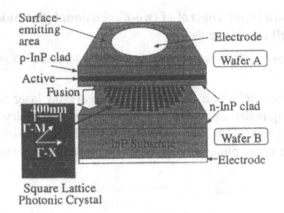

Figure 5.2.2(4-3) Schematic structure of fabricated 2D photonic crystal laser.

(5) "Directional lasing oscillation of two-dimensional organic photonic crystal lasers at several photonic band gaps"

M. Notomi, H. Suzuki, and T. Tamamura (NTT, Japan), Appl. Phys. Lett. **78**, 1325 (2001).

A 2D triangular lattice photonic crystal was used as a laser cavity, which was photopumped using an N_2 laser operating at 337 nm. The authors showed that it is possible to produce lasing at the M, K and Γ points, simply by changing the pitch and the angle of the photonic crystal.

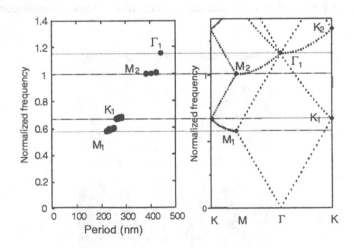

Figure 5.2.2(5-1) Measured normalized frequencies at which the lasing oscillation was observed and the photonic band diagram.

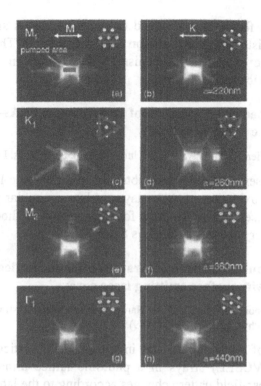

Figure 5.2.2(5-2) Various pictures showing the lasing oscillation.

(6) "Recurrent-photon feedback in two-dimensional photonic-crystal lasers"

S. Nojima (NTT, Japan), Phys. Rev. B **65**, 073103 (2002).

A 2D photonic crystal laser of finite size was analyzed. A new type of

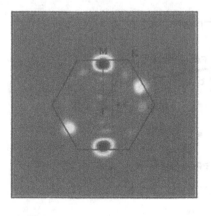
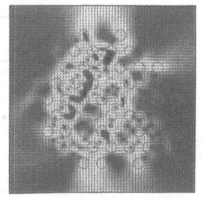

Figure 5.2.2(6) Field intensity distribution of a reciprocal lattice space (left), and photon flux (Poynting power) distribution (right).

recurrent-photon feedback was found to occur through a simulation of the field intensity distribution and photon flux distribution. The feedback was explained by the expanded mechanism which occurs in an ordinary 1D distributed feedback (DFB) laser.

(7) "Photopumped laser operation of an oxide post GaAs-AlAs superlattice photonic lattice"

P. W. Evans, J. J. Wierer, and N. Holonyak (Univ. Illinois), Appl. Phys. Lett. **70**, 1119 (1997).

Photopumped laser oscillation was obtained in a photonic lattice fabricated by the periodic oxidation of AlAs layer. Also, a circular pillar crystal in which Zn is diffused as a substitution for an oxidation method was fabricated. The pitch of the crystal was as large as 8 μm.

(8) "Modal expansion analysis of strained photonic lattices based on vertical cavity surface emitting laser arrays"

T. Fishman, E. Kapon, H. Pier, and A. Hardy (Swiss Federal Institute of Technology, Switzerland), Appl. Phys. Lett. **74**, 3595 (1999).

The possibility of mode control was investigated for vertical cavity surface emitting laser (VCSEL) arrays in a photonic lattice arrangement. It was found that the near-field pattern changes according to the lattice arrangement.

(9) "Photonic crystal laser mediated by polaritons"

Nojima (NTT, Japan), Phys. Rev. B **61**, 9940 (2000).

A photonic crystal laser utilizing polaritons was proposed. The threshold gain of this type of laser is found to be sufficiently low, compared with other lasers. It is reported that the laser can oscillate in a single mode due to strong mode selectivity.

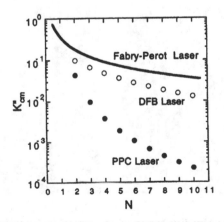

Figure 5.2.2(9) Comparison of threshold gains among different three-type lasers.

5.3 LIGHT-EMITTING DIODES

An effective refractive index of a photonic crystal could be lower than those of conventional materials such as semiconductors when the structure is carefully designed. A light extraction efficiency could be increased by utilizing this property of photonic crystal. The surface recombination should be avoided to increase the efficiency in photonic crystal based light-emitting diodes (LEDs).

(1) "High extraction efficiency of spontaneous emission from slabs of photonic crystals"

S. Fan, P. R. Villeneuve, and J. D. Joannopoulos (MIT, USA), Phys. Rev. Lett. **78**, 3294 (1997).

The results calculating the direction of spontaneous emission generated in a 2D photonic crystal slab with a circular hole are discussed. It is expected that over 80% of the light generated would be emitted in the vertical direction as a result of suppression of the light propagation in the in-plane direction, due to the in-plane PBG effect.

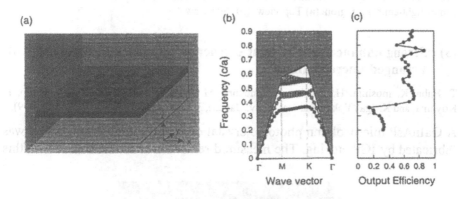

Figure 5.3(1) Calculation model, band diagram, and the light extraction efficiency.

(2) "Light extraction from optically pumped light-emitting diode by thin-slab photonic crystals"

M. Boroditsky, T. F. Krauss, R. Coccioli, R. Vrijen, R. Bhat, and E. Yablonovitch (UCLA, USA), Appl. Phys. Lett. **75**, 1036 (1999).

A 2D photonic crystal with circular holes was formed on a semiconductor slab. It was then bonded to a glass substrate, and an epitaxial lift-off was performed to fabricate a thin slab photonic crystal. The photoluminescence (PL) generated at a wide center region without holes was guided in the slab and radiated in the upper direction by the hole arrangement surrounding the center region. The light extraction efficiency was increased by more than a factor of ten when changes were made to the crystal pitch. The sample showing the strongest PL intensity is estimated to possess a light extraction efficiency of over 70%.

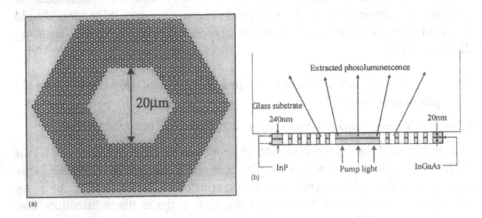

Figure 5.3(2) Schematic structure of fabricated semiconductor 2D photonic crystal slab with center light-emitting region. (a) Top view. (b) Side view.

(3) "Strong enhancement of light extraction efficiency in GaInAsP 2-D-arranged microcolumns"

T. Baba, K. Inoshita, H. Tanaka, J. Yonekura, M. Ariga, A. Matsutani, T. Miyamoto, F. Koyama, and K. Iga (Yokohama Nat. Univ., Japan), J. Lightwave Technol. **17**, 2113 (1999).

A GaInAsP microcolumn photonic crystal of a honeycomb arrangement was fabricated by ICP etching. The measured result for emission lifetimes in this

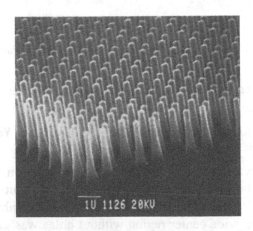

Figure 5.3(3-1) GaInAsP circular column photonic crystal.

crystal show an emission intensity increased by more than ten times that for a broad wafer. This crystal may be utilized for fabrication of a high efficiency LED. However, the lowering of internal efficiency due to surface recombination at the sidewall of the microcolumn was found to be a serious issue.

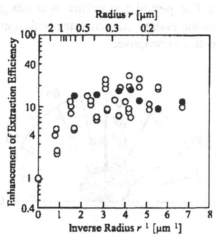

Figure 5.3(3-2) Increases in the light extraction efficiency as a function of the circular column radius.

(4) "Surface recombination measurements on III-V candidate materials for nanostructure light-emitting diodes"

M. Boroditsky, I. Gontijo, M Jackson, R. Vrijen, E. Yablonovitch, T. Krauss, C. C. Cheng, A. Scherer, R. Bhat, and M. Krames (UCLA, etc., USA), J. Appl. Phys. **87**, 3497 (2000).

The surface recombination velocity of AlGaN, InGaAs, and InGaAlP were investigated by measuring the quantum efficiency of photoluminescence. This analysis revealed that InGaAs and GaN were found to be the most appropriate materials for the fabrication of photonic crystal LEDs.

(5) "Rate-equation analysis of output efficiency and modulation rate of photonic-crystal Light-emitting diodes"

S. Fan, P. R. Villeneuve, and J. D. Joannopoulos (MIT, USA), IEEE J. Quantum Electron. **36**, 1123 (2000).

The characteristics of a photonic crystal were analyzed. Nonradiative recombination and photon reabsorption processes are incorporated into the rate-equation for investigation.

(6) "Highly directive light sources using two-dimensional photonic crystal slabs"

A. L. Fehrembach, S. Enoch, and A. Sentenac (Univ. St. Jerome, France), Appl. Phys. Lett. **79**, 4280 (2001).

A microcavity with a fine periodic structure was designed. A structure in which 80% of all radiation energy concentrates within an angle of 0.2° from the perpendicular direction is proposed.

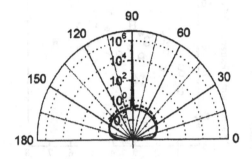

Figure 5.3(6) Radiation pattern.

(7) "Reduction in surface recombination of GaInAsP microcolumns by CH₄ plasma irradiation"

H. Ichikawa, K. Inoshita, and T. Baba (Yokohama Nat. Univ., Japan), Appl. Phys. Lett. **78**, 2119 (2001).

The surface recombination velocity in microcolumns of GaInAsP for 1.55 μm was evaluated through a carrier lifetime measurement using a phase resolved spectroscopy. It was reported that CH_4 ECR plasma irradiation decreased the velocity by one half and increased the PL intensity to 2 – 5 times.

5.4 RESONATORS AND FILTERS

The following papers describe the utilization of artificial defects in PBG for the realization of ultrasmall functional components and/or devices such as resonators (cavities) and filters. Theoretical calculations were first carried out mainly for pure 2D photonic crystal with infinite height. After various tries and errors, the first clear experimental demonstration of such ultrasmall devices composed of line-defect waveguide and point-defect was achieved in 2000.

(1) "Microcavities in photonic crystals: mode symmetry, tenability, and coupling efficiency"

P. R. Villeneuve, S. Fan, and J. D. Joannopoulos (MIT, USA), Phys. Rev. Lett. **54**, 7837 (1996).

The resonant Q-value generated by the introduction of defects in a 2D photonic crystal of circular column arrangement is calculated. The Q-value is found to be over 10 000 when the crystal size is 9×9 periods.

"Channel drop tunneling through localized states"

S. Fan, P. R. Villeneuve, J. D. Joannopoulos, and H. A. Haus (MIT, USA), Phys. Rev. Lett. **80**, 960 (1996).

"Theoretical analysis of channel drop tunneling processes"

S. Fan, P. R. Villeneuve, J. D. Joannopoulos, M. J. Khan, C. Manolatou, and H. A. Haus (MIT, USA), Phys. Rev. B **59**, 15882 (1999).

The analysis of a device made of a photonic crystal incorporating defects sandwiched between parallel waveguides, is presented. Such a device may be used as a resonant filter, or to provide light coupling between waveguides.

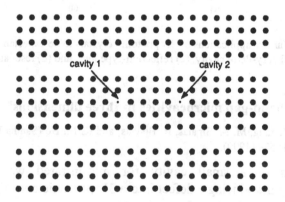

Figure 5.4(1) Point defect cavity in a 2D photonic crystal

(2) "Demonstration of cavity mode between two-dimensional photonic-crystal mirrors"

D. Labilloy, H. Benisty, C. Weisbuch, T. F. Krauss, V. Bardinal, and U. Oesterle (École Polytechnique, France, etc.), Electron. Lett. **33**, 1978 (1997).

A resonant mode at a wavelength of 0.94 μm was observed in a cavity formed by a pair of 2D photonic crystals which were placed in parallel with a small separation.

"Optical and confinement properties of two-dimensional photonic crystals"

H. Benisty, C. Weisbuch, D. Labilloy, M. Rattier, C. J. M. Smith, T. F. Krauss, R. M. De La Rue, R. Houdre, U. Oesterle, C. Jouanin, and D. Cassagne (Ecole Polytechnique, France), J. Lightwave Technol. **17**, 2063 (1999).

A 2D photonic crystal of circular holes with concentric cylinder structures was fabricated. In the introductory chapter, various application examples were explained from a general point of view.

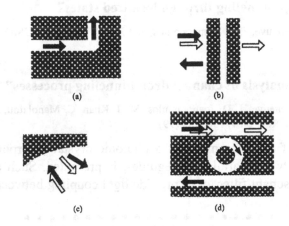

Figure 5.4(2) Various applications of 2D photonic crystals to be expected, (a) waveguide with bends, (b) Fabry-Perot cavity, (c) reflection-type lens, and (d) resonant filter.

(3) "Photonic band-gap microcavities in three dimensions"

S. Y. Lin, J. G. Fleming, M. M. Sigalas, R. Biswas, and K. M. Ho (Sandia Nat. Lab., USA), Phys. Rev. B **59**, 15579 (1999).

The formation of a small cavity for mid-infrared wavelengths was investigated, by introducing a defect into a 3D photonic crystal. Single mode resonance in the mid-infrared wavelength region was confirmed by the measurement of transmission and reflection characteristics.

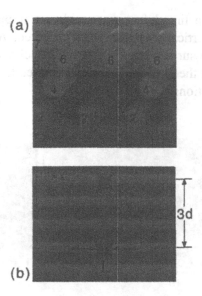

Figure 5.4(3-1) SEM image of the fabricated sample.

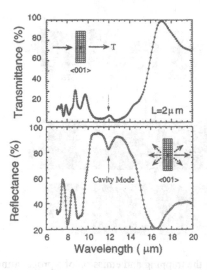

Figure 5.4(3-2) Transmission and reflection spectra.

(4) "Trapping and emission of photons by single defect in a photonic bandgap structure"

S. Noda, A. Chutinan, and M. Imada (Kyoto Univ., Japan), Nature **407**, 608 (2000).

This is the first clear experimental demonstration of a single defect filtering device in the optical frequency regime. The phenomena such that photons propagating through a line-defect waveguide are trapped by a point-defect and emitted to the vertical direction were observed. A suitable application for such a device is a surface-emitting type channel add/drop devices. Also, the author claims that the phenomena can be applied to the trapping of atoms and nonlinear applications.

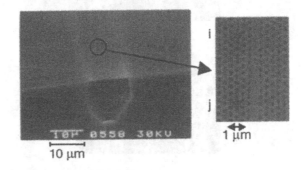

Figure 5.4(4-1) SEM image of the fabricated sample.

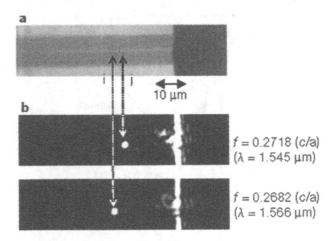

Figure 5.4(4-2) Image of the trapping and emission of a propagating photon in a waveguide.

(5) "Directionally dependent confinement in photonic-crystal microcavities"

C. J. M. Smith, T. F. Krauss, H. Benisty, M. Rattier, C. Weisbuch, U. Oesterle, and R. Houdre (Univ. Grasgow, etc., UK), J. Opt. Soc. Am. B **17**, 2043 (2000).

Characteristics of various cavities in triangular lattice 2D photonic crystals on GaAs/AlGaAs substrates were evaluated. Ten different defect patterns were investigated. Over 95% reflectivity at the mirror was reported despite the strong directional dependence. As light confinement effect for vertical direction was weak in their structure, a relatively large defect size was introduced in order to obtain a large Q-value.

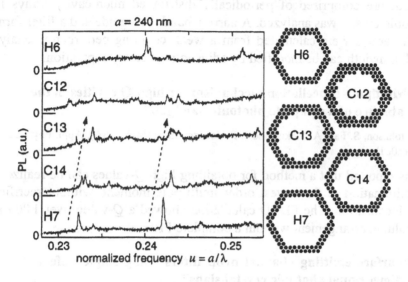

Figure 5.4(5) Various cavity structures and their resonance spectra.

(6) "Quality factor for localized defect modes in a photonic crystal slab upon a low-index dielectric substrate"

E. Miyai, and K. Sakoda (Hokkaido Univ., Japan), Opt. Lett. **26**, 740 (2001).

The localized defect mode of a 2D photonic crystal slab on a SiO_2 substrate was calculated numerically by the FDTD method. A degenerate mode with E_1 symmetry was found to exist in the pseudo-gap of the slab with a Q-value of 800.

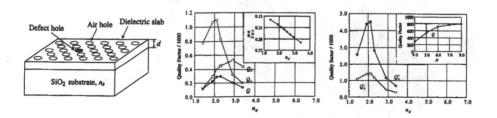

Figure 5.4(6) Model of a defect-cavity in a 2D photonic crystal on SiO_2 substrate and its characteristics.

(7) "Narrow-band microcavity waveguides in photonic crystals"

A. Boag, and B. Z. Steinberg (Tel Aviv Univ., Israel), J. Opt. Soc. Am. A **18**, 2799 (2001).

A structure comprised of periodically distributed microcavity arrays in a photonic crystal was analyzed. A narrow band waveguide and a filter formed by the arrays were calculated from a weak coupling perturbation analysis. The bandwidth can be modified by adjusting the cavity separation.

(8) "Multipole-cancellation mechanism for high-Q cavities in the absence of a complete photonic bandgap"

S. G. Johnson, S. Fan, A. Mekis, and J. D. Joannopoulos (MIT, USA), Appl. Phys. Lett. **78**, 3388 (2001).

It was reported that a method for obtaining high Q-values can be realized by the elimination of the lowest order multi-pole moment, without sacrificing light localization. The FDTD calculation showed a Q-value over 1 000 in a 2D column arrangement without a complete PBG.

(9) "Surface-emitting channel drop filters using single defects in two-dimensional photonic crystal slabs"

A. Chutinan, M. Mochizuki, M. Imada, and S. Noda (Kyoto Univ., Japan), Appl. Phys. Lett. **79**, 2690 (2001).

This paper describes the detailed calculation of the paper (4). The FDTD calculation was used to analyze the efficiency of trapping and emission of photons by the single defect at the vicinity of the photonic crystal waveguide. It was shown that the appropriate selection of defect arrangement enables tuning of operation wavelength, keeping the radiation efficiency maximum.

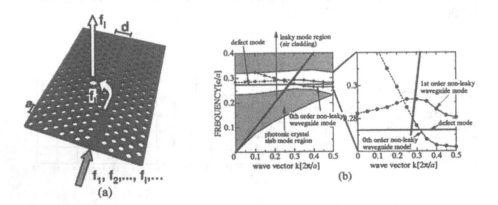

Figure 5.4(9) Schematic structure of a channel drop filter using a line-defect waveguide and a point-defect cavity in 2D photonic crystal slab (a), and the corresponding band diagram (b).

(10) "Direct spectroscopy of a deep two-dimensional photonic crystal microresonator"

P. Kramper, A. Birner, M. Agio, C. M. Soukoulis, F. Muller, U. Gosele, J. Mlynek, and V. Sandoghdar (Univ. Konstanz, etc., Germany), Phys. Rev. B **64**, 233102, (2001).

A photonic crystal made of porous silicon was fabricated with a PBG from 3.4 – 5.8 μm. Two resonant peaks agreeing well with the numerically calculated results were observed in the measured spectra of a small 2D photonic crystal cavity which possesses such a point defect.

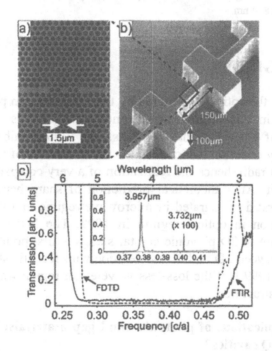

Figure 5.4(10) SEM images along with the observed transmission and resonant spectra of the fabricated device.

(11) "Design of photonic crystal microcavities for cavity QED"

J. Vuckovic, M. Loncar, H. Mabuchi, and A. Scherer (Caltech, USA), Phys. Rev. E **65**, 016608 (2002).

The optimization of a microcavity in a 2D photonic crystal slab was investigated for possible enhancement of the coupling intensity between the cavity field and a single defect cavity. Numerical calculation results regarding the Q-value and the volume of the localized defect mode are presented as a function of geometric parameters.

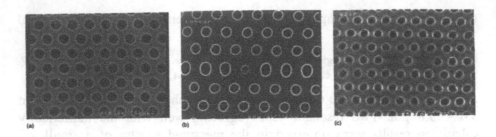

Figure 5.4(11) SEM images of the fabricated photonic crystals. It has a microcavity operating at a wavelength of 854 nm.

5.5 WAVEGUIDES

Application for the optical waveguide by the PBG effect in photonic crystals are discussed in this section. A channel waveguide is realized by the incorporation of a linear chain of defects. The PBG prohibits light leakage into an arbitrary direction, which leads to the realization of waveguides with ultrasmall bend radii, hence the realization of a very compact optical and/or photonic circuit. After theoretical prediction of sharp bend waveguide in 1996, it was first demonstrated in microwave regime in 1998, followed by the demonstration in optical regime in 1999. The loss-less condition in realistic structure of 2D photonic crystal slab was first pointed out in 2000. Comparison between 2D and 3D waveguide with realistic structures, it was pointed out in 1999 that the loss-less wavelength region was much broader in 3D than 2D structures.

(1) "Novel applications of photonic band gap materials: low-loss bends and high Q cavities"

R. D. Meade, A. Devenyi, J. D. Joannopoulos, O. L. Alerhand, D. A. Smith, and K. Kash (MIT, USA), J. Appl. Phys. **75**, 4753 (1994).

A laser cavity and a waveguide are discussed, where the 2D photonic crystal is considered to be a mirror and a clad, respectively.

"High transmission through sharp bends in photonic crystal waveguides"

A. Mekis, J. C. Chen, I. Kurland, S. Fan, P. R. Villeneuve, and J. D. Joannopoulos (MIT, USA), Phys. Rev. Lett. **77**, 3787 (1996).

A right angle bent waveguide using a 2D photonic crystal clad of a square lattice structure was analyzed by the 2D FDTD method. Transmission of over 95% was estimated for a wide frequency range.

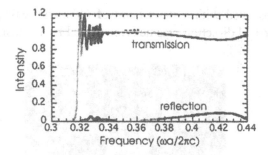

Figure 5.5(1-1) Transmission and reflection characteristics of a right angle bent waveguide.

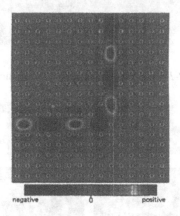

Figure 5.5(1-2) Light propagation results at the bend.

"Experimental demonstration of guiding and bending of electromagnetic waves in a photonic crystal"

S. Y. Lin, E. Chow, V. Hietala, P. R. Villeneuve, and J. D. Joannopoulos (MIT, USA), Science **282**, 274 (1998).

Sharp bend waveguide in 2D photonic crystal with enough height for third direction was demonstrated in microwave regime. Almost 100%transmission was experimentally confirmed. The radius of curvature was designed for one wavelength.

(2) "Observation of light propagation in photonic crystal optical waveguides with bends"

T. Baba, N. Fukaya, and J. Yonekura (Yokohama Nat. Univ., Japan), Electron. Lett. **35**, 654 (1999).

A triangular lattice photonic crystal slab was formed by bonding a semiconductor thin film on an SiO$_2$. Propagation of light at a wavelength of

1.5 μm was observed with a steep bend of 60°. The leakage of the guided mode in the vertical direction was not considered in the design.

Figure 5.5(2-1) Image of the fabricated waveguide pattern.

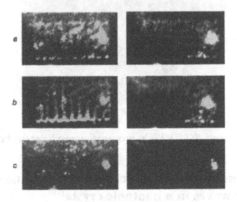

Figure 5.5(2-2) Near-field images of the observed propagating light.

"Analysis of finite 2-D photonic crystals of columns and lightwave devices using the scattering matrix method"

J. Yonekura, M. Ikeda, and T. Baba (Yokohama Nat. Univ., Japan), J. Lightwave Technol. **17**, 1500 (1999).

A 2D photonic crystal optical waveguide was analyzed by the scattering matrix method. A waveguide having bends of 60° and 120° in a triangular lattice arrangement, a Y-splitter of 120° and a very short directional coupler were discussed. It was shown that high transmission can be obtained by a fine control of elements at bends and waveguide ends.

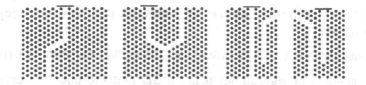

Figure 5.5(2-3) Various waveguide structures considered in calculations.

(3) "Highly confined waveguides and waveguide bends in three-dimensional photonic crystal"

A. Chutinan, and S. Noda (Kyoto Univ., Japan), Appl. Phys. Lett. **75**, 3739 (1999)

A proposal for a waveguide device with a steep bend, fabricated from a 3D photonic crystal is analyzed. This is a 3D structure with stacked stripes, however the parallel rods are stacked with a shift of half a period. A right angle bent waveguide is realized by removing one rod from each direction. Transmission of over 95% is demonstrated for a wide frequency range that fully covers the fiber communication wavelength range.

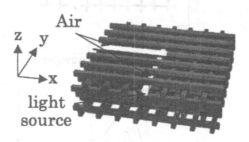

Figure. 5.5(3-1) Right angle bent waveguide in 3D photonic crystal.

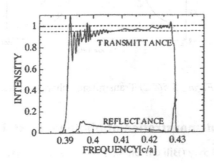

Figure. 5.5(3-2) Transmission and reflection spectra.

(4) "Waveguides in three-dimensional metallic photonic band-gap materials"

M. Soukoulis, and D. D. Crouch (Iowa State Univ., USA), Phys. Rev. B **60**, 4426 (1999).

A right angle bent waveguides using a 3D metallic photonic crystal was analyzed. A 3D mesh was assumed to be burying a dielectric having a refractive index of 1.5. An approximation was made of the structure, allowing analysis by the FDTD method to be possible. The structure comprised of a defect inserted into the intermediate layer of a waveguide composed of $10 \times 10 \times 3$ wires in the x, y and z directions respectively. The measured transmission was found to be 85%.

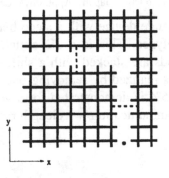

Figure 5.5(4-1) L-type waveguide

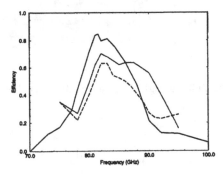

Figure 5.5(4-2) Transmission characteristics.

(5) "Experimental demonstration of photonic crystal based waveguides"

B. Temelkuran, and E. Ozbay (Bilkent Univ., Turkey), Appl. Phys. Lett. **74**, 486 (1999).

Microwave transmission characteristics for a waveguide with a rectangular bend was measured for a 3D photonic crystal into which alumina rods stacked layer-by-layer were introduced. Transmission at around 10 – 13 GHz band was confirmed.

(6) "Band gap and wave guiding effect in a quasiperiodic photonic crystal"

C. Jin, B. Cheng, B. Man, Z. Li, D. Zhang, S. Ban, and B. Sun (Chinese Academy of Science, China), Appl. Phys. Lett. **75**, 1848 (1999).

Transmission characteristics for microwaves in a 2D photonic crystal fabricated from alumina rods were investigated. The pattern in the crystal had a quasi-periodicity. The introduction of an optical waveguide into such a crystal with a complete PBG, leads to favorable transmission characteristics.

(7) "Waveguide bend in three-dimensional layer-by-layer photonic bandgap materials"

M. M. Sigalas, R. Biswas, K. M. Ho, C. M. Soukoulis, D. Turner, B. Vasiliu, S. C. Kothari, and S. Lin (Iowa State Univ., etc., USA), Microw. Opt. Technol. Lett. **23**, 56 (1999).

A simulation by the FDTD method was made for a waveguide, formed by the removal of rods from a 3D photonic crystal of stacked alumina rods. Transmission of 100% can be obtained for a rectangular bend, depending on how the rods are removed.

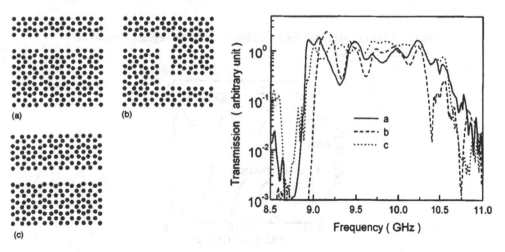

Figure 5.5(7) Optical waveguide composed of a quasiperiodic 2D photonic crystal, and the observed transmission spectra.

(8) "Waveguides and waveguide bends in two-dimensional photonic crystal slabs"

A. Chutinan, and S. Noda (Kyoto Univ., Japan), Phys. Rev. B **62**, 4488 (2000).

A straight waveguide and a waveguide bend are analyzed for a photonic crystal slab with air cladding for vertical direction (airbridge-type waveguide). A guideline for the design of these waveguides without any leakage loss is given.

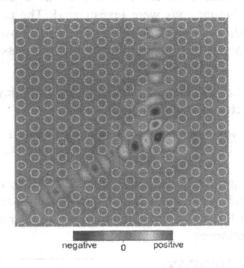

negative   0   positive

Figure 5.5(8-1) Calculated field distribution at the waveguide bend in 2D photonic crystal slab.

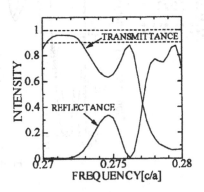

Figure 5.5(8-2) Transmission and reflection spectra for 60° bend.

(9) "Duplexer using microwave photonic band gap structure"

S. S. Oh, C. S. Kee, J. E. Kim, H. Y. Park, T. I. Kim, I. Park, and H. Lim (KAIST, etc., Korea), Appl. Phys. Lett. **76**, 2301 (2000).

A duplexer (a transmitting and receiving splitter) which shows frequency selectivity in the microwave regime was designed by utilizing two photonic crystals with different PBGs, and its characteristics were calculated.

(10) "Two-dimensional photonic crystal waveguides with 60° bends in a thin slab structure"

N. Fukaya, D. Ohsaki, and T. Baba (Yokohama Nat. Univ., Japan), Jpn. J. Appl. Phys. **39**, 2619 (2000).

The research on the line defect waveguide in a photonic crystal slab, which was reported in paper (2), was described in detail. Light propagation in the straight and bent waveguides was observed in the 1.51 – 1.55 µm range.

(11) "Waveguiding in planar photonic crystals"

M. Loncar, D. Nedeljkovic, T. Doll, J. Vuckovic, A. Scherer, and T. P. Pearsall (Caltech, USA), Appl. Phys. Lett. **77**, 1937 (2000).

A 2D photonic crystal waveguide on a silicon-on-insulator (SOI) substrate was designed and fabricated. Propagation of guided light was observed at a wavelength around 1.55 µm even in 60° bend.

"Design and fabrication of silicon photonic crystal optical waveguides"

Same group, J. Lightwave Technol. **18**, 1402 (2000).

The design and fabrication of a 2D photonic crystal were reported in detail. Photonic bands for the guided mode of the standard and modified line defect waveguide structures were calculated by a 3D-FDTD calculation. The waveguides including 60° bend were fabricated on an SOI substrate.

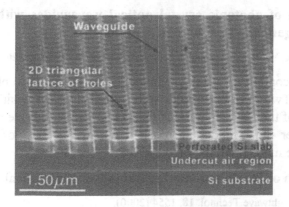

Figure 5.5(11-1) SEM picture of waveguide in airbridge-type photonic crystal slab.

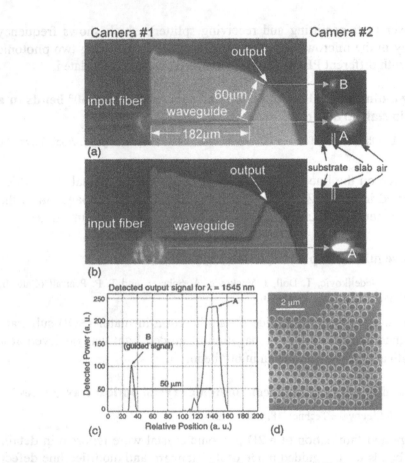

Figure 5.5(11-2) Near field patterns of a photonic crystal waveguide incorporating 60° bends (a) and (b), the output ratio of emitted light (c), and SEM image of the waveguide (d).

(12) "Design of photonic crystal optical waveguides with single mode propagation in the photonic bandgap"

A. Adibi, R. K. Lee, Y. Xu, A. Yariv, and A. Scherer (Caltech, USA), Electron. Lett. **36**, 1376 (2000).

Single mode conditions were systematically investigated in photonic crystal waveguides. Two lines of airholes were introduced; the positioning of these on the sides of the waveguide remained fixed, whilst the size of the hole was varied. Higher order modes are pushed out of the PBG of this structure, making single mode propagation feasible.

"Propagation of the slab modes in photonic crystal optical waveguides"

Same group, J. Lightwave Technol. **18**, 1554 (2000).

The photonic crystal waveguide was theoretically calculated and discussed. The propagation mode in this waveguide can be controlled so that it resembles the mode in an ordinary dielectric waveguide. The coupling of the waveguide and an ordinary waveguide can be realized with a high efficiency, using such a controlled mode.

(13) "Optical demultiplexing in a planar waveguide with colloidal crystal"

I. Avrutsky, V. Kochergin, and Y. Zhao (Wayne State Univ., USA), IEEE Photon. Technol. Lett. **12**, 1647 (2000).

Wavelength splitting was shown in the 1.55 μm range by adding a 2D grating to a plane waveguide. This grating was formed on the waveguide by stacking one layer of a colloidal crystal, which self-assembles into a hexagonal arrangement.

(14) "Radiation losses of waveguide-based two-dimensional photonic crystals: positive role of the substrate"

H. Benisty, D. Labilloy, C. Weisbuch, C. J. M. Smith, T. F. Krauss, D. Cassagne, A. Beraud, and C. Jouanin (Ecole Polytechnique, etc., France), Appl. Phys. Lett. **76**, 532 (2000).

A photonic crystal fabricated by the etching of a plane waveguide, was analyzed using first order perturbation theory, to calculate the radiation loss. The loss decreases when the cladding refractive index is designed to be nearly equal to the core refractive index. This result agrees with experimental result for the near infrared wavelength range.

(15) "Two-dimensional vector-coupled-mode theory for textured planar waveguides"

P. Paddon, and J. F. Young (Univ. British Columbia, Canada), Phys. Rev. B **61**, 2090 (2000).

A model to treat the coupling between guided modes in a planar dielectric waveguide is developed based on a general Green's function technique, which self-consistently determines the electric field distribution on the surface 2D photonic crystal. Light in the crystal is represented as a linear combination of eigenmodes. Various unique characteristics that depend on the phase relation are obtained, including polarization characteristics and the lifetime of leaky modes.

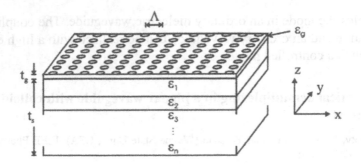

Figure. 5.5(15) Typical model of the sample used.

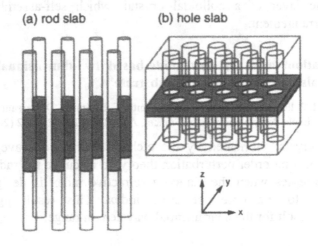

Figure 5.5(16-1) Schematics of photonic crystal slab.

(16) "Linear waveguides in photonic-crystal slabs"

S. G. Johnson, P. R. Villeneuve, S. Fan, and J. D. Joannopoulos (MIT, USA), Phys. Rev. B **62**, 8212 (2001).

Line defect waveguides in photonic crystal slabs and in cylinder arrays are numerically analyzed. The results are different from those for 2D crystals with an infinite height. Structures needed for single mode propagation are investigated by changing various parameters.

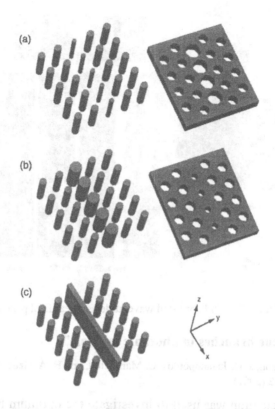

Figure 5.5(16-2) Various line defects in photonic crystal slabs.

(17) "Clear correspondence between theoretical and experimental light propagation characteristics in photonic crystal waveguides"

T. Baba, N. Fukaya, and A. Motegi (Yokohama Nat. Univ., Japan), Electron. Lett. **37**, 761 (2001).

A waveguiding loss of 11 dB/mm was evaluated for a fabricated single line defect waveguide in an airbridge photonic crystal slab. Both transmission bands and near-field patterns, agree well with the calculated results of the photonic band.

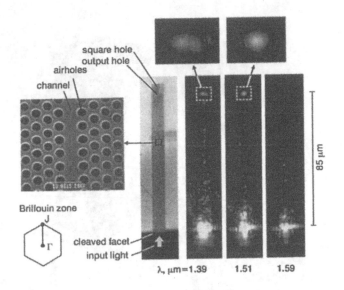

Figure 5.5(17) Fabricated waveguide and near field pattern.

(18) "Waveguide branches in photonic crystals"

S. Fan, S. G. Johnson, J. D. Joannopoulos, C. Manolatou, and H. A. Haus (MIT, USA), J. Opt. Soc. Am. B **18**, 162 (2001).

The FDTD calculation was used to investigate the optimum branch structure in a photonic crystal waveguide for the highest transmission efficiency and the lowest reflection. The transmission is increased by the addition of small columns at the T-branch.

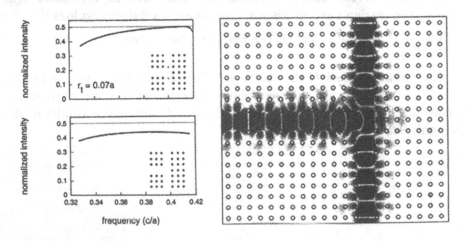

Figure 5.5(18) Transmission spectra at an ordinary T-branch (lower left) and that of a T-branch with smaller columns added (upper left). The simulated electric field distribution for the optimized branch (right).

(19) "Transmission and time of flight study of $Al_xGa_{1-x}As$-based photonic crystal waveguides"

N. Kawai, K. Inoue, N. Ikeda, N. Carlsson, Y. Sugimoto, K. Asakawa, S. Yamada, and Y. Katayama (Hokkaido Univ., etc., Japan), Phys. Rev. B **63**, 153313, (2001).

Light propagation characteristics are investigated for a photonic crystal slab fabricated on an AlGaAs substrate. Light radiation in the upward direction is significantly increased with an increase in the air filling factor in the slab. The condition to suppress this radiation is reported to be the filling factor of less then 0.35.

"Confined band gap in an air-bridge type of two-dimensional AlGaAs photonic crystal"

Same group, Phys. Rev. Lett. **86**, 2289 (2001).

A transmission spectrum measured for an airbridge photonic crystal slab made of AlGaAs agreed with calculated band curves. Existence of a waveguiding mode and a PBG for the TE-like mode were expected.

(20) "Role of distributed Bragg reflection in photonic-crystal optical waveguides"

A. Adibi, Y. Xu, R. K. Lee, M. Loncar, A. Yariv, and A. Scherer (Caltech., etc., USA), Phys. Rev. B **64**, 041102, (2001).

High-accuracy mode calculation in a photonic crystal waveguide is expected to be possible by transforming the photonic crystal clad to an equivalent grating structure around the center slab. Such a transformation is useful not only for the analysis and review of a combination of different waveguides, but also for designing PBG waveguides.

(21) "Bend loss in surface plasmon polariton band-gap structures"

S. I. Bozhevolnyi, V. S. Volkov, K. Leosson, and A. Boltasseva (Aalborg Univ., etc., Denmark), Appl. Phys. Lett. **79**, 1076 (2001).

"Waveguiding in surface plasmon polariton band gap structures"

Same group, Phys. Rev. Lett. **86**, 3008 (2001).

The propagation of a surface plasmon polariton was investigated by near-field optical microscopy. A 2 μm wide line defect photonic crystal waveguide with bends on an Au surface was used as a sample. The bend losses for 15° and 30° bends were 2 dB and 10 dB, respectively. Similar characteristics for Y-branches were also investigated.

(22) "Polymer photonic crystal slab waveguides"

C. Liguda, G. Bottger, A. Kuligk, R. Blum, M. Eich, H. Roth, J. Kunert, W. Morgenroth, H. Elsner, and H. G. Meyer (Univ. Hambrug-Harburg etc., Germany), Appl. Phys. Lett. **78**, 2434 (2001).

A 2D photonic crystal waveguide of BCB polymer on a Teflon substrate was designed, fabricated and evaluated. Circular holes of 300 nm in diameter in a square lattice of 500 nm pitch were fabricated by EB lithography and RIE. Measurement for TE-like and TM-like modes at 1.3 μm were in good agreement with FDTD calculations. The TE-like mode suppression for a 10 layer photonic crystal at a PBG was 15 dB.

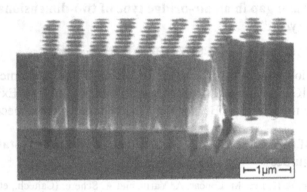

Figure 5.5(22) Cross-sectional SEM image. The sample was etched up to the Teflon substrate.

(23) "Design of impurity band-based photonic crystal waveguides and delay lines for ultra short optical pulse"

S. Lan, S. Nishikawa, H. Ishikawa, and O. Wada (FESTA, Japan), J. Appl. Phys. **90**, 4321 (2001).

The transmission of an extremely short pulse in a photonic crystal waveguide with an impurity band was examined. In general, the transmission properties are strongly affected by the pulse width corresponding to the resonant spectral width of the impurity band. A careful design of the waveguide structure allows a flat impurity band.

(24) "Out-of-plane losses of two-dimensional photonic crystal waveguides: electromagnetic analysis"

P. Lalanne, and H. Benisty (CNRS, etc., France), J. Appl. Phys. **89**, 1512 (2001).

An analysis for the radiation loss in photonic crystal waveguides is presented. The results agree quantitatively with experimental results for AlGaAs slab 2D photonic crystals.

(25) "Observation of light propagation in two-dimensional photonic crystal-based bent optical waveguides"

S. Yamada, T. Koyama, Y. Katayama, N. Ikeda, Y. Sugimoto, K. Asakawa, N. Kawai, and K. Inoue (Tsukuba Univ., etc., Japan), J. Appl. Phys. **89**, 855 (2001).

The light propagation at near infrared wavelengths is observed in a AlGaAs/GaAs photonic crystal waveguide with bends. The wavelength range that exhibits light propagation corresponds well to the PBG.

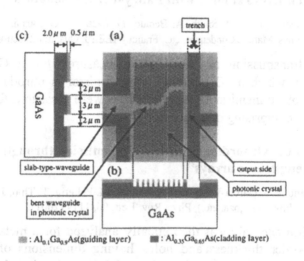

Figure. 5.5(25) Typical model of the sample used in this paper.

(26) "Interference of signals in parallel waveguides in a two-dimensional photonic crystal"

M. Qui, and S. He (Dept. Electromag. Theory etc., Sweden), Physica B **299**, 187 (2001).

A coupling between parallel waveguides in a 2D photonic crystal composed of dielectric rods was calculated. Two cases were considered; one was the insertion of two linear waveguides, and the other was lines defect waveguides. The former is advantageous for the optical confinement, so suppresses the interaction by one line interval. The latter requires five line interval to obtain sufficient suppression.

(27) "Improved line-defect structures for photonic crystal waveguides with high group velocity"

K. Yamada, H. Morita, A. Shinya, and M. Notomi (NTT, Japan), Opt. Commun. **198**, 395 (2001)

Various modified line defect waveguides in a 2D photonic crystal were analyzed. The dispersion curve of the guided mode is explained as a mutual interaction between an index-confined mode and a PBG-confined mode. The result indicated the importance of a large group velocity near the zone boundary, and the management of the light line problem for a photonic crystal slab sandwiched by low index dielectrics.

(28) "Performance of waveguide-based two-dimensional photonic-crystal mirrors studied with Fabry-Perot resonators"

M. Rattier, H. Benisty, C. J. M. Smith, A. Beraud, D. Cassagne, C. Jouanin, T. F. Krauss, C. Weisbuch (Lab. Phys. Matiere Condensee, etc., France), IEEE J. Quantum Electron. **37**, 237 (2001).

Reflection, transmission and loss were measured for a GaAs-AlGaAs waveguide, in which a 2D photonic crystal mirror is introduced. An 88% reflection, a 6% transmission and a 6% loss were measured for a photonic crystal mirror comprising of 4 periods.

(29) "Theory of extraordinary optical transmission through sub-wavelength hole arrays"

L. Martin-Moreno, F. J. Garcia-Vidal, H. J. Lezec, K. M. Pellerin, T. Thio, J. B. Pendry, and T. W. Ebbesen (Univ. Zaragoza, etc.), Phys. Rev. Lett. **86**, 1114 (2001).

A transmission anomaly was theoretically analyzed for a metal film with a wavelength order thickness and holes having dimensions of less than a wavelength. An optimum analytical model was developed to show an enhancement in transmission due to the tunneling of surface plasmons generated at the metal dielectric interface.

(30) "Optical trirefringence in photonic crystal waveguides"

M. C. Netti, A. Harris, and J. J. Baumberg (Univ. Southampton, etc., UK), Phys. Rev. Lett. **86**, 1526 (2001).

The existence of three refractive indices is demonstrated in a 2D photonic crystal. The indices maximally have six axes. This property in not observed in any birefringent dielectrics. A photonic crystal waveguide with a Fabry-Perot cavity was then fabricated to demonstrate this phenomenon experimentally.

(31) "Singlemode transmission within photonic bandgap of width-varied single-line-defect photonic crystal waveguides on SOI substrates"

M. Notomi, K. Yamada, A. Shinya, J. Takahashi, C. Takahashi, and I. Yokohama (NTT, Japan), Electron. Lett. **37**, 293 (2001).

A single mode propagation was realized by controlling the width of a single line defect waveguide fabricated on an SOI substrate. The propagation loss of 6 dB/mm is reported.

"Extremely large group-velocity dispersion of line-defect waveguides in photonic crystal slabs"

Same group, Phys. Rev. Lett. **87**, 253902 (2001).

Light propagation characteristics for line defect waveguides in a photonic

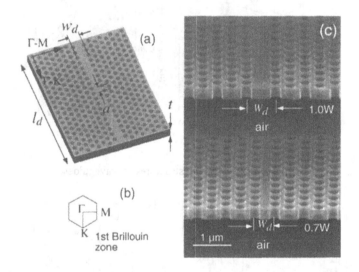

Figure 5.5(31-1) Schematic and SEM views of the waveguide structure.

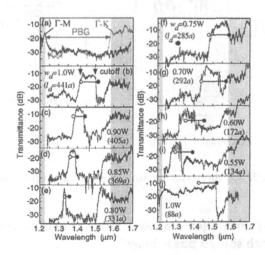

Figure 5.5(31-2) Transmission spectra of waveguides with different defect widths.

crystal slab were measured and compared with the dispersion relations derived by the 3D FDTD calculation. It was shown from the Fabry-Perot resonance in the waveguide that the group refractive index near the zone boundary is far higher than the refractive index of the medium.

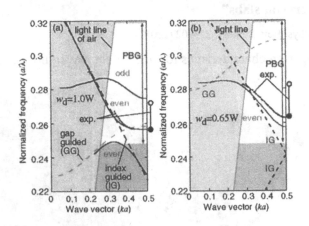

Figure 5.5(31-3) Band structures of waveguides.

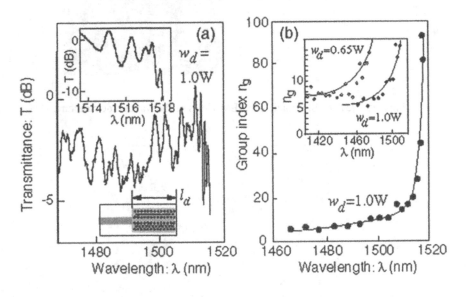

Figure 5.5(31-4) Fabry-Perot resonance spectra and group refractive index.

"Singlemode transmission in SOI-type photonic-crystal line-defect waveguides with shifted phase holes"

Same group, Electron. Lett. **38**, 74 (2002).

A calculation and an experiment were carried out for a waveguide where the airholes at the core was shifted. A steep bend cannot be realized in width-modulated waveguides due to the break of symmetry in the photonic crystal of the bent waveguide, whilst this problem is solved in the hole-shifted waveguide. A good single mode propagation can be obtained for this type of waveguide even on an SOI substrate.

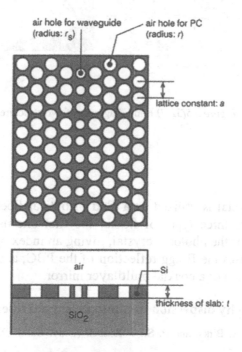

Figure 5.5(31-5) Model of the hole-shifted waveguide.

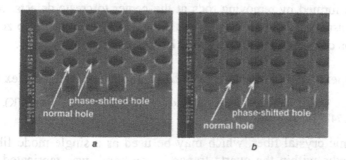

Figure 5.5(31-6) SEM views of fabricated structures. Center holes in the left image are smaller than others.

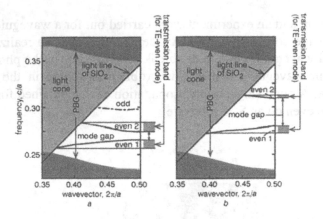

Figure 5.5(32-7) Band diagrams of two structures.

5.6 FIBERS

The photonic crystal is utilized as a clad of optical fibers in the following papers. There are three types of fibers; the first one uses a low effective refractive index of the photonic crystal, giving an index optical confinement, the second one uses the Bragg reflection of the PBG, and the third one uses the Bragg reflection of a coaxial multilayer mirror.

(1) "Group-velocity dispersion in photonic crystal fibers"

D. Mogilevtsev, T. A. Birks, and P. St. J. Russell (Univ. Bath, UK), Opt. Lett. **23**, 1662 (1998).

A triangular lattice 2D photonic crystal was fabricated in an optical fiber. A fiber core was formed by removing rods at the center to create defects. It is shown that the dispersion at wavelengths less than 1.27 µm can be set to zero and the dispersion compensation is possible at 1.55 µm by suitable designs.

(2) "Properties of photonic crystal fiber and the effective index model"

J. C. Knight, T. A. Birks, P. St. J. Russell, and J. P. de Sandro (Univ. Bath, UK), J. Opt. Soc. Am. A **15**, 748 (1998).

A photonic crystal fiber, which may be used as a single mode fiber for all wavelengths within the quartz transmission band, was fabricated. A single mode is observed experimentally in the wavelength range from 337 nm to over 1550 nm. It is shown from a calculation of equivalent refractive index that fiber dimensions have no relation to the single mode condition.

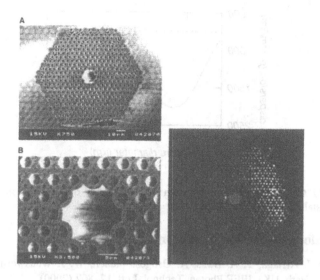

Figure 5.6(2) Cross-section of a photonic crystal fiber and a near-field image of a visible beam passing through it.

(3) "A photonic crystal fiber"

P. Rigby (Univ. Oxford), Nature **396**, 415 (1998).

Photonic crystal fibers accomplished at Univ. of Bath are reviewed. Fabrication results are shown by SEM images.

(4) "Single-mode photonic band gap guidance of light in air"

R. F. Cregan, B. J. Mangan, J. C. Knight, T. A. Birks, P. S. Russell, P. J. Roberts, and D. C. Allan (Univ. Bath, UK), Science **285**, 1537 (1999).

A large airhole is used as the fiber core, which is surrounded by photonic crystal clad made from silica tubes. Light propagation in a singlemode is demonstrated. As nonlinear effects are suppressed even at high power in the air core, this fiber is expected to be applicable to a large scale ultrahigh density wavelength division multiplexing.

(5) "Dispersion compensation using single-material fibers"

T. A. Birks, D. Mogilevtsev, J. C. Knight, and P. St. J. Russell (Univ. Bath, UK), IEEE Photon. Technol. Lett. **11**, 674 (1999).

Research into the dispersion properties of photonic crystal fibers was calculated. It was shown that a negative coefficient of dispersion over 1000 times larger than that of an ordinary single mode fiber is achievable. This could be effective for low loss compact dispersion compensation.

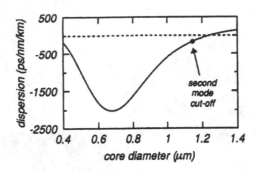

Figure 5.6(5) Calculated relation between core diameter and dispersion coefficient for photonic crystal fibers.

(6) Anomalous dispersion in photonic crystal fiber

J. C. Knight, J. Arriaga, T. A. Birks, A. Ortigosa-blanch, W. J. Wadsworth, and P. St. J. Russell (Univ. Bath, UK), IEEE Photon. Technol. Lett. **12**, 807 (2000).

Group velocity dispersion characteristics of a silica photonic crystal fiber were measured in both the visible and near infrared wavelength ranges. A single mode fiber with zero dispersion at a wavelength of 700 nm was demonstrated. This fiber will play an important role in the generation of soliton pulses.

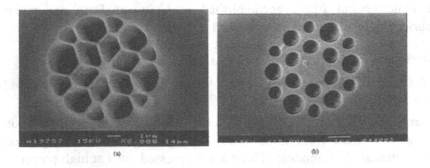

Figure 5.6(6) Cross-sectional SEM image of a photonic crystal fiber. (a) Core diameter of 1 μm. (b) Core diameter of 1.5 μm and an airhole diameter of 0.62 μm.

(7) "Soliton effects in photonic crystal fibres at 850 nm"

W. J. Wadsworth, J. C. Knight, A. Ortigosa-Blanch, J. Arriaga, E. Silvestre, and P. St. J. Russell (Univ. Bath, UK), Electron. Lett. **36**, 53 (2000).

Unique group velocity properties and soliton effects, which are never observed in ordinary fibers, were observed in a silica photonic crystal fiber at a wavelength of 850 nm. Zero group velocity dispersion at 740 nm was also demonstrated.

(8) "The guidance properties of multi-core photonic crystal fibres"

P. J. Roberts, and T. J. Shepherd (QinetiQ, Malvern, UK), J. Opt. A **3**, S133 (2001).

The propagation characteristics of a photonic crystal fiber possessing three cores were calculated. Degenerate propagation modes were found when interaction between the cores are considered.

(9) "Design of weakly guiding Bragg fibres for chromatic dispersion shifting towards short wavelengths"

B. J. Mangan, J. Arriaga, T. A. Birks, J. C. Knight, and P. St. J. Russell (Univ. Bath, UK), Opt. Lett. **26**, 1469 (2001).

Propagation characteristics in a fiber with a multilayer coaxial Bargg grating were calculated. It becomes possible to set a dispersion-less region in the 1.3 μm wavelength range by the appropriate selection of refractive index distribution.

(10) "Fundamental-mode cutoff in a photonic crystal fiber with a depressed-index core"

B. J. Mangan, J. Arriaga, T. A. Birks, J. C. Knight, and P. St. J. Russell (Univ. Bath, UK), Opt. Lett. **26**, 1469 (2001)

A holey fiber whose core has a lower refractive index due to fluorine doping, is presented. The effective refractive indices of the core and clad coincide with each other at a certain wavelength, which is at the lower limit of the light propagation range of the fiber.

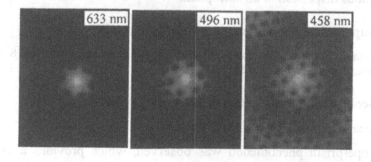

Figure 5.6(10-1) Near-field images at the fiber end.

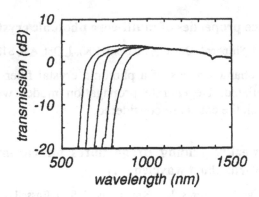

Figure 5.6(10-2) Transmission spectra for various types of bend.

(11) "Photonic crystal fibers: an endless variety"

T. A. Birks, J. C. Knight, B. J. Mangan, and P. St. J. Russell (Univ. Bath, UK), IEICE Trans. Electron. **E84-C**, 585 (2001).

Theories and experiments on photonic crystal fibers are reviewed. Its similarity and difference with conventional optical fibers are discussed.

5.7 PRISMS AND POLARIZERS

Papers discussed here propose the utilization of photonic crystals mainly for bulk devices. In particular, there are many proposals for the utilization of anomalous dispersion or anisotropism.

(1) "Superprism phenomena in photonic crystals"

H. Kosaka, T. Kawashima, A. Tomita, M. Notomi, T. Tamamura, T. Sato, and S. Kawakami (NEC, etc., Japan), Phys. Rev. B **58**, 10096 (1998).

"Self-collimating phenomena in photonic crystals"

Same group, Appl. Phys. Lett. **74**, 1212 (1999).

The superprism phenomenon was observed, which provides a very large angular dispersion of over 100 times that in an ordinary prism. It was experimentally demonstrated in a Si/SiO_2 3D photonic crystal and discussed with the dispersion surface of photonic bands. Divergence and collimation of light were also demonstrated and explained by the same theory. A small optical circuit was suggested as an application of these phenomena.

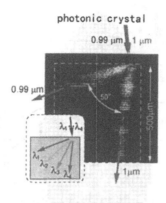

Figure 5.7(1-1) Observation of the superprism phenomenon.

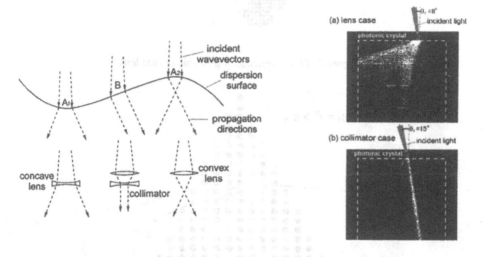

Figure 5.7(1-2) Schematic of dispersion curve and light propagation direction in a photonic crystal (left), and actual observations from samples (right).

(2) "Photonic crystals as optical components"

P. Halevi, A. A. Krokhin, and J. Arriaga (Optica y Electronica, Mexico), Appl. Phys. Lett. **75**, 2725 (1999).

"Photonic crystal optics and homogenization of 2d periodic composites"
Same group, Phys. Rev. Lett. **82**, 719 (1999).

Brirefringent optical components utilizing a frequency region below the PBG was proposed. A polarization-dependent lens was given as an example application. In addition, 2D photonic crystals were discussed as either mono or biaxial structure. It is easy to design many kinds of optical devices by use of simple equations for the effective dielectric constant.

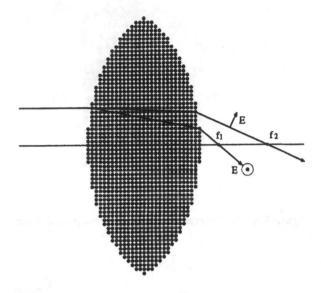

Figure 5.7(2-1) Birefringent photonic crystal lens.

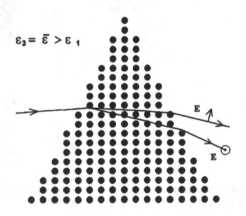

Figure 5.7(2-2) Photonic crystal prism which can be designed with the proposed method.

(3) "Photonic crystal polarization splitters"

Y. Ohtera, T. Sato, T. Kawashima, T. Tamamura, and S. Kawakami (Tohoku Univ., Japan), Electron. Lett. **35**, 1271 (1999).

The Si/SiO$_2$ multilayer was formed by bias sputtering technique on a substrate with many stripe-like steps to be a kind of 2D photonic crystal. The multilayer with a *zigzag* type cross-section works as a polarization splitter, which can be used for light incidence from normal direction to the surface. An insertion loss of 0.4 dB and an extinction ratio of 40 dB were realized at a wavelength of 1.5 µm. A device for the visible wavelength range was realizable by changing the material to TiO_2/SiO_2.

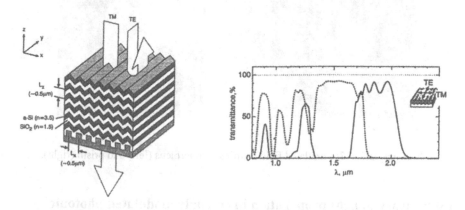

Figure 5.7(3) Structure of a polarization splitter (left), and observed transmission spectra for each polarization (right).

(4) "Anomalous refractive properties of photonic crystals"

B. Gralak, S. Enoch, and Gerard Tayeb (Centre de Saint-Jerome, France), J. Opt. Soc. Am. A **17**, 1012, (2000).

Light propagation in a 2D photonic crystal was investigated using the scattering matrix method. Various devices such as negative refractive gratings, microlenses and microprisms were simulated.

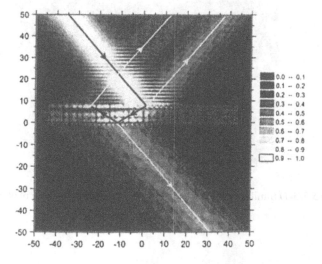

Figure 5.7(4-1) Simulated field profile of negative refraction of light.

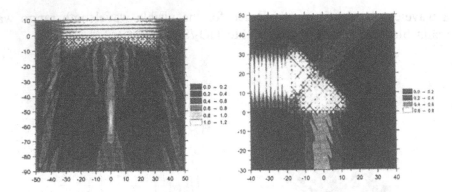

Figure 5.7(4-2) Simulated field profiles of microlens (left) and prism (right).

(5) "Theory of light propagation in strongly modulated photonic crystals: refraction-like behavior in the vicinity of the photonic band gap"

M. Notomi (NTT, Japan), Phys. Rev. B. **62**, 10696 (2000).

It is argued that a strong modulation of refractive index in photonic crystals provides a negative refractive phenomenon, which can be understood by Snell's law with a negative index value. This phenomenon opens up some interesting application possibilities such as a unique imaging system, which has never been realized with conventional optics.

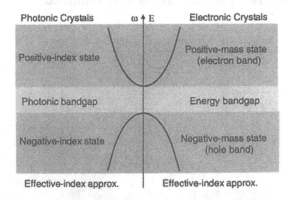

Figure 5.7(5-1) Similarity of concepts between an electronic and a photonic system.

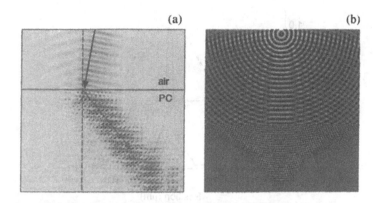

Figure 5.7(5-2) Light propagation characteristics of anomalous refractive indices.

(6) "Photonic-crystal spot-size converter"

H. Kosaka, T. Kawashima, A. Tomita, T. Sato, and S. Kawakami (NEC, Japan), Appl. Phys. Lett. **76**, 268 (2000).

The possibility of a spotsize converter using a photonic crystal is discussed theoretically and experimentally. A conversion efficiency of 10:1 was estimated. In comparison with a conventional spotsize converter, the advantages of this device are its small size, independence from incident position, and deep focusing depth.

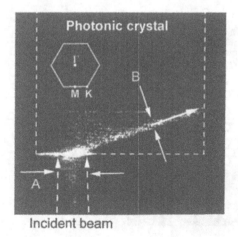

Figure 5.7(6-1) Observed light transmission in a photonic crystal, showing the beam narrowing.

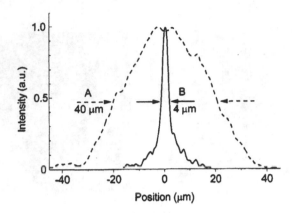

Figure 5.7(6-2) Light beam profile before and after the focusing.

(7) "Interfaces of photonic crystals for high efficiency light transmission"

T. Baba, and D. Ohsaki (Yokohama Nat. Univ., Japan), Jpn. J. Appl. Phys. **40**, 5920 (2001).

The improvement of transmission efficiency in a 2D photonic crystal at a higher frequency than the PBG was theoretically investigated. An increase in the efficiency was confirmed by the FDTD analysis for the cases when smaller elements or projection-type elements are placed at in-out edge. In regard to the projection-type element, it is possible to increase the maximum efficiency to almost 100%.

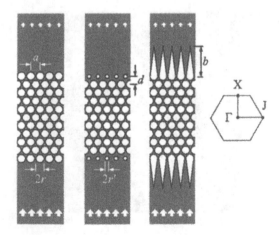

Figure 5.7(7) Calculated structure for light transmission. Basic structures (left), smaller element structure (center), and projection-type element structure (right).

(8) "Superprism effect in opal-based photonic crystals"

T. Ochiai, and J. Sanchez-Dehesa (Univ. Autónoma de Madrid, Spain), Phys. Rev. B **64**, 245113 (2001).

Superprism phenomenon in 3D opal photonic crystals is discussed. This phenomenon is rather found in a frequency range which contains many flat bands. Numerical calculations suggested the possibility for their future application to a beam splitter and a monochrometer.

5.8 PHOTONIC INTEGRATED CIRCUITS

The following papers deal with ultrasmall and high-density photonic integrated circuits that combine light emitters, waveguides and functional devices all based on the photonic crystal concept.

(1) "Photonic crystals: putting a new twist on light"

J. D. Joannopoulos, P. R. Villeneuve, and S. Fan (MIT, USA), Nature **386**, 143 (1997).

The feasibility of putting sharp bends in photonic crystal waveguides is reported. The key areas of discussion are the calculation results along with information on the use of 1D photonic crystals as waveguides and cavities. Their illustration is adopted in a cover page of this article to explain future images of photonic integrated circuits combined with various photonic crystal elements.

(2) "Microlaser and photonic integrated circuit by photonic crystals"

T. Baba (Yokohama Nat. Univ., Japan), IEICE **81**, 1067 (1998, in Japanese)

The integration of waveguide-type elements and devices such as bends, splitters or directional couplers by use of photonic crystals in a very small area were suggested. The proposed WDM circuit requires only 1/1000 of the area ordinarily required.

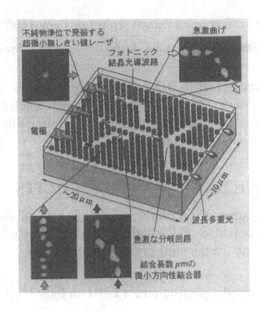

Figure 5.8(2) Schematic of a photonic circuit composed of point and line defects formed in a 2D photonic crystal.

(3) "Alignment and stacking of semiconductor photonic bandgaps by wafer-fusion"

S. Noda, N. Yamamoto, M. Imada, H. Kobayashi, and M. Okano (Kyoto Univ., Japan), J. Lightwave Technol. **17**, 1948 (1999).

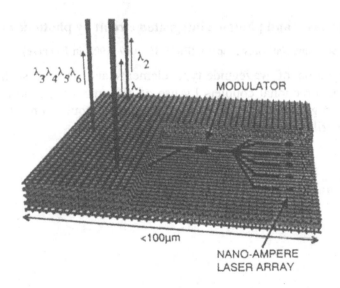

Figure 5.8(3) Photonic integrated circuits utilizing complete PBG 3D photonic crystals.

A 3D photonic IC, which integrates various functional devices such as laser arrays, waveguides with sharp bends, modulators, and wavelength selective functional devices in an ultrasmall area of less than 100 μm^2 were proposed. They are to be fabricated by the introduction of defects or light emitters in a complete PBG 3D photonic crystal. The feasibility of fabricating these ICs by use of the wafer fusion technique is shown.

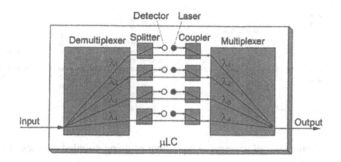

Figure 5.8(4) Schematic of add-drop WDM circuit composed of photonic crystals and superprisms.

(4) "Photonic crystals for micro lightwave circuits using wavelength-dependent angular beam steeling"

H. Kosaka, T. Kawashima, A. Tomita, M. Notomi, T. Tamamura, T. Sato, and S. Kawakami (NEC, Japan), Appl. Phys. Lett. **74**, 1370 (1999).

A combination of more than one photonic crystal, which show superprism phenomena, was proposed to form small optical circuits. In particular, add-drop devices and dispersion compensators for WDM systems are indicated to be possible in a very small area.

(5) "Photonic crystal circuits: a theory for two- and three-dimensional networks"

A. R. McGurn (Western Michigan Univ., USA), Phys. Rev. B **61**, 13235 (2000).

"Photonic crystal circuits: localized modes and waveguide couplers"

Same group, Phys. Rev. B **65**, 075406 (2002).

A waveguide network composed of coupled cavity channels is discussed. Various components such as bends, splitters, crosses, couplers, and so on are described.

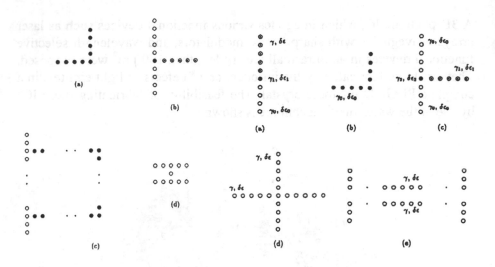

Figure 5.8(5) Schematic of waveguide type devices using coupled cavities in a square lattice 2D photonic crystal.

(6) "Fabrication and characterization of different types of two-dimensional AlGaAs photonic crystal slabs"

Y. Sugimoto, N. Ikeda, N. Carlsson, K. Asakawa, N. Kawai, and K. Inoue (FESTA, etc., Japan), J. Appl. Phys. **61**, 922 (2002).

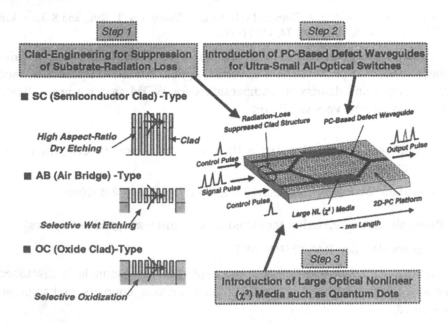

Figure 5.8(6) Schematic of a photonic crystal based Mach-Zehnder optical switch.

A Mach-Zehnder type all optical switch with a large $\chi^{(3)}$ material is described with line defect waveguides in a photonic crystal slab. In the experiment, three kinds of photonic crystal slab having a semiconductor clad, an airbridge and an oxide clad were fabricated and evaluated.

5.9 NONLINEAR DEVICES

The nonlinear properties and dispersion relations in photonic crystals, particularly at band-edges, are discussed in the following papers.

(1) "Photonic bound states in periodic dielectric materials"

R. D. Meade, K. D. Brommer, A. M. Rappe, and J. D. Joannopoulos (MIT, USA), Phys. Rev. B **44**, 13772 (1991).

Modes generated in the PBG due to defects are reviewed. Spherical defects of dielectrics and air are introduced in crystals composed of dielectric spheres and a diamond structure. A high Q cavity and a tunable wavelength filter are listed as feasible devices.

(2) "Sum-frequency generation in a two-dimensional photonic lattice"

K. Sakoda, and K. Ohtaka (Hokkaido Univ., Japan), Phys. Rev. B **54**, 5742 (1996).

The theory of higher-order frequency generation in photonic crystals is discussed. Several enhancements are pointed out in regard to the effects of zero group velocity after the calculation of phase-matching conditions.

(3) "Second harmonic generation in a photonic crystal"

J. Martorell, R. Vilaseca, and R. Corbalan (Univ. Autonoma de Barcelona, Spain), Appl. Phys. Lett. **70**, 702 (1997).

It is shown that phase matched second harmonic generation (SHG) is possible using of photonic crystals. An fcc photonic crystal, composed of polystyrene spheres of 0.115 μm in diameter and coated with a dye was prepared. A 532 nm optical pulse was obtained from 1064 nm laser pulse irradiation. SHG is demonstrated to occur in the vicinity of the PBG by a numerical analysis. Phase matching occurs at the band-edge.

(4) "Nonlinear photonic crystals"

V. Berger (Thomson CSD Laboratoire, France), Phys. Rev. Lett. **81**, 4136 (1998).

2D photonic crystals with $\chi^{(2)}$ nonlinearity were considered theoretically. Crystals which possess frequency dependencies of the linear and nonlinear susceptibilities can be generalized by 1D pseudo-phase-matched structures, which are feasible to make using LiNbO$_3$ and GaAs.

(5) "Propagation and second-harmonic generation of electromagnetic waves in a coupled-resonator optical waveguide"

Y. Xu, R. K. Lee, and A. Yariv (Caltech, USA), J. Opt. Soc. Am. B **17**, 387 (2000).

The tight-binding method and the FDTD method were applied to analyze two kinds of coupled-cavity optical waveguide (CROW) structures. Dispersion relations of these structures were determined by the small coupling coefficient. The spatial characteristics of the modes are found to be similar to modes of single cavity with high Q-values. A significant application for this would be to cross-talk-less waveguides and an increase in the efficiency of SHG.

(6) "Raman gap solitons"

H. G. Winful, and V. Perlin (Univ. Michigan, USA), Phys. Rev. Lett. **84**, 3586 (2000).

Strong pumping with high power pulses, where the wavelength is far from Bragg resonances in a nonlinear periodic structure, results in the excitation of gap solitons at a wavelength which coincides with the Raman-shift of the PBG material. At this Raman gap, solitons are stable and have a long lifetime, and they can almost remain stationary in one place. This phenomenon is predicted to occur in nonlinear PBG structures.

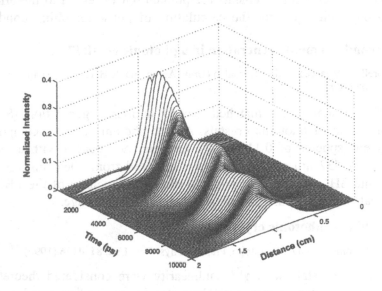

Figure 5.9(6) Time dependant expansion of solitons.

(7) "Hexagonally poled lithium niobate: a two-dimensional nonlinear photonic crystal"

N. G. R. Broderick, G. W. Ross, H. L. Offerhaus, D. J. Richardson, and D. C. Hanna (Univ. Southampton, UK), Phys. Rev. Lett. **84**, 4345 (2000).

Nonlinear 2D photonic crystals were fabricated whose second order nonlinear susceptibility was periodic in space despite their constant refractive index. An SHG wave, which is pseudo-phase-matched, appears in such a crystal. The efficiency was measured to be over 60% with the use of picosecond pulses.

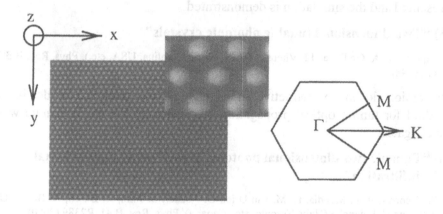

Figure 5.9(7) Upper side image of a fabricated HeXLN crystal and its Brillouin zone.

5.10 TUNABLE CRYSTALS AND OPTICAL SWITCHES

Changes in band structure which are accompanied with a refractive index change are studied in papers of this section.

(1) "Single-mode waveguide microcavity for fast optical switching"

P. R. Villeneuve, D. S. Abrams, S. Fan, and J. D. Joannopoulos (MIT, USA), Opt. Lett. **21**, 2017 (1996).

The change of refractive index in an airbridge cavity by the laser light irradiation was proposed and analyzed as a method for realizing modulators and switches by changing a resonant mode in the PBG. Ionization of deep donors is shown to possibly provide a larger change of the refractive index than ordinary methods like the applied voltage or current injection.

(2) "Optical switching with a nonlinear photonic crystal: a numerical study"

P. Tran (Naval Air Warfare Center Weapons Div., USA), Opt. Lett. **21**, 1138 (1996).

"Optical limiting and switching of short pulses by use of a nonlinear photonic bandgap structure with a defect"

Same group, J. Opt. Soc. Am. B **14**, 2589 (1997).

"All-optical switching with a nonlinear chiral photonic bandgap structure"

Same group, J. Opt. Soc. Am. B **16**, 70 (1999).

The FDTD method for photonic crystals with the third order nonlinearity is presented and the simulation is demonstrated.

(3) "Two-dimensional tunable photonic crystals"

A. Figotin, Y. A. Godin, and I. Vitebsky (Univ. North Carolina, USA, etc.), Phys. Rev. B **57**, 2841 (1998).

Dispersion dependent refractive index change was analyzed in detail. A method for tuning optical propagation properties by the slight change was investigated.

(4) "Tunable two-dimensional photonic crystal using liquid-crystal infiltration"

S. W. Leonard, J. P. Mondia, H. M. van Driel, O. Toader, S. John, K. Busch, A. Birner, U. Gosele, and V. Lehmann (Univ. Toronto, etc., Canada), Phys. Rev. B **41**, R2389 (2000).

The PBG in a 2D photonic crystal was altered by use of liquid crystal. The refractive index of a liquid crystal is temperature dependent. A 2D photonic crystal of 1.58 μm pitch was filled with liquid crystal to confirm changes in the PBG from 3.3 – 5.7 μm to 4.4 – 6.0 μm after filling with liquid crystal.

Figure 5.10(4-1) Si 2D photonic crystal fabricated by the porous etching technique.

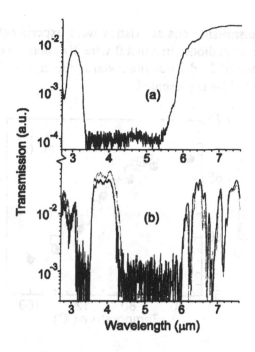

Figure 5.10(4-2) Changes of PBG. (a) Before liquid crystal filling. (b) After filling. Solid lines and dotted lines in (b) denote 35°C, and 62°C respectively.

(5) "Experimental demonstration of electrically controllable photonic crystal at centimeter wavelength"

A. de Lustrac, F. Gadot, S. Cabaret, J. M. Lourtioz, T. Brillat, A. Priou, and E. Akmansoy (Univ. Paris, France), Appl. Phys. Lett. **75**, 1625 (1999).

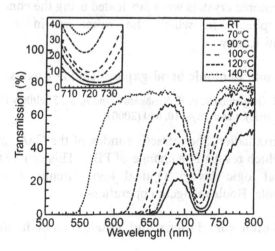

Figure 5.10(5-1) Temperature characteristics of transmission spectra.

The PBG and transmission characteristics were experimentally changed by the electric field on pin diodes in a metal wire 2D photonic crystal. A change in the transmittance of 25 dB was observed in the crystal. The transmission change according to bias is presented.

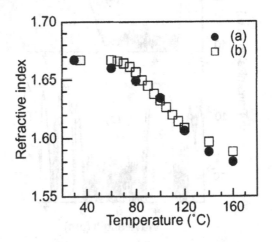

Figure 5.10(5-2) Temperature dependence of refractive indices. (a) Direct measurement. (b) Estimations from stopbands.

(6) "Tunable optical stop band utilizing thermochromism of synthetic opal infiltrated with conducting polymer"

S. Satoh, H. Kajii, Y. Kawagishi, A. Fujii, M. Ozaki, and K. Yoshino (Osaka Univ., Japan), Jpn. J. Appl. Phys. **38**, L1475 (1999).

Tunable opal photonic crystals were fabricated using the conducting polymer poly-3-Alkylthiophene, in which the refractive index changed with temperature significantly.

(7) "Low-threshold photonic band-gap optical logic gates"

I. S. Nefedov, V. N. Gusyatnikov, P. K. Kashkarov, and A. M. Zheltikov (Inst. of Radio Eng. & Electron., etc., Russia), Laser Phys. **10**, 640 (2000).

By using light irradiation, the refractive index of the GaAs photonic crystal was changed, which results in a change of PBG. This can be used for a low-threshold optical logic device. Optical logic circuits were fabricated to realize fundamental Boolean algebra operations.

(8) "Photon-photon correlations and entanglement in doped photonic crystals"

D. Petrisyan, and G. Kurizki (Weizmann Institute of Science Phys., Israel), Phys. Rev. A **64**, 023810 (2001).

This paper considers the situation that four-level atomic transitions are resonating and coupled with the edge of the PBG in a 2D photonic crystal. The strong phase or amplitude correlation appears in the atomic emission. One photon induces a large phase shift or causes absorption triggering for the other photon. Such a system could work as an ultrahigh-speed nonlinear optical switch.

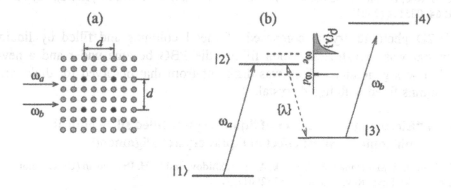

Figure 5.10(8) (a) 2D photonic crystals with low concentration of doping. Black points represent doped atoms. (b) Four-level atomic transition.

(9) "Tunable photonic crystal with semiconductor constituents"

P. Halevi, and F. R. Mendieta (Inst. Nacional de Astrofisica, Mexico), Phys. Rev. Lett. **85**, 1875 (2000).

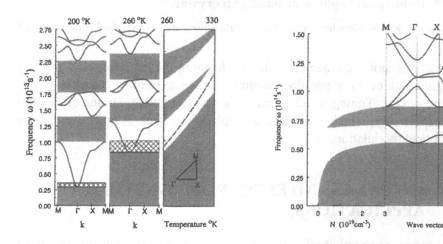

Figure 5.10(9) The change in photonic bands. Temperature dependence (left) and impurity concentration dependence (right).

The photonic band of semiconductor photonic crystals is thought to be variable under conditions of high enough carrier density. Dielectric constants strongly depend on temperature and plasma frequency, which can be changed by the impurity concentration. A PBG can be eliminated under these conditions.

(10) "Two-dimensional tunable metallic photonic crystals infiltrated with liquid crystals"

C. S. Kee, H. Lim, Y. K. Ha, J. E. Kim, and H. Y. Park (Ajou Univ., etc., Korea), Phys. Rev. B **64**, 085114 (2001).

A 2D photonic crystal composed of metal columns and filled by liquid crystal was investigated. Upon filling, the PBG became wider and a new PBG was generated, which was different from that observed for dielectric columns filled with liquid crystal.

(11) "Electro-optic behavior of liquid-crystal-filled silica optical photonic crystals: effect of liquid crystal alignment"

D. Kan, J. E. Maclennan, N. A. Clark, A. A. Zakhidov, and R. H. Baughman (Univ. Colorado, etc., USA), Phys. Rev. Lett. **86**, 4052 (2001).

Photonic crystals composed of silica spheres filled with nematic liquid crystal were investigated. When the sphere surface was parallel to the molecular axis of the liquid crystal, the change of Bragg reflection due to an electric field was enhanced. This effect did not appear when the molecular axis was perpendicular to the sphere surface.

(12) "Strain-tunable photonic band gap crystals"

K. Sungwon, and V. Gopalan (Pennsylvania State Univ., etc., USA), Appl. Phys. Lett. **78**, 3015 (2001).

Calculations indicated the possibility of changing the PBG from the ordinary triangular lattice to a pseudo triangular lattice system by using a piezo-electric device. Tuning of 52% and 73% was possible at strains of 2% and 3% respectively, which could lead to applications such as optical switching and optical modulation.

5.11 ANTENNAS AND ELECTROMAGNETIC WAVE APPLICATIONS

The attempts to use photonic crystals for radio frequencies are covered in the following papers. For example, it may be possible to increase antenna gain by use of a coupling of radio waves to leaky modes of photonic crystal substrate.

(1) "Radiation properties of a planar antenna on a photonic-crystal substrate"

E. R. Brown, C. D. Parker, and E. Yablonovitch (MIT, etc., USA), J. Opt. Soc. Am. B 10, 404 (1993).

Photonic crystal characteristics were investigated, when used as substrate material for microwave and millimeter wave planar antennas. The antenna was confirmed to radiate waves out to the free space rather than into the substrate at 13.2 GHz, when a photonic crystal with a PBG from 13 – 16 GHz was used.

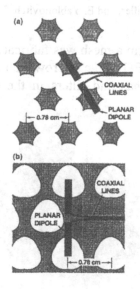

Figure 5.11(1) Structure of a dipole antenna combined with a photonic crystal.

"Effect of surface composition on the radiation pattern from a photonic-crystal planar-dipole antenna"

Same group, Appl. Phys. Lett. 64, 3345 (1994).

Radiation patterns of a planar dipole antenna composed of a photonic crystal with a center PBG frequency of 14.5 GHz were measured. Radiation patterns were critically dependent upon the filling factor in the photonic crystal, becoming narrow as the factor decreased. A circular hole antenna showed stronger directivity.

"Photonic crystals: a new quasi-optical component for high-power microwaves"

K. Agi, L. D. Moreland, E. Schamiloglu, M. Mojahedi, K. J. Malloy, and E. R. Brown (Univ. New Mexico, etc., USA), IEEE Trans. Plasma, **24**, 1067 (1996).

Interference between a fcc photonic crystal and high power microwaves have been studied. Airholes periodically arranged in the dielectric of the photonic crystal exhibited optical characteristics similar to metal mirrors. Antennas and filters are discussed as possible applications high power microwaves.

(2) "3D wire mesh photonic crystals"

D. F. Sievenpiper, M. E. Sickmiller, and E. Yablonovitch (UCLA, USA), Phys. Rev. Lett. **76**, 2480 (1995).

Metal wires stacked to form a mesh was fabricated as a 3D photonic crystal, which can have a wide PBG in the microwave range. This could be utilized in applications for antennas and filters in the microwave and millimeter wave range.

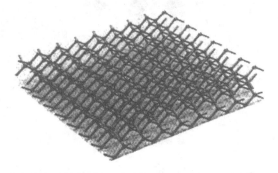

Figure 5.11(2) Fabricated metal wire mesh.

(3) "Optimized dipole antennas on photonic band gap crystals"

S. D. Cheng, R. Biswas, E. Ozbay, S. McCalmont, G. Tuttle, and K. M. Ho (Iowa State Univ., USA), Appl. Phys. Lett. **67**, 3399 (1995).

Complete reflection from a photonic crystal was utilized to fabricate a planar dipole antenna with high efficiency. It exhibited optimum operation in terms of directionality and driving frequency.

(4) "Photonic band-gap materials for high-gain printed circuit antennas"

H. Y. D. Yang, N. G. Alexopoulos, and E. Yablonovitch (Univ. Illinois, etc., USA), IEEE Trans. Antennas Propag. **45**, 185 (1997).

The gain increase in certain directions was theoretically demonstrated for a printed circuit antenna in combination with a 2D photonic crystal. Space waves, bound (surface) waves, and leaky waves exist in the photonic crystal. Of these, surface waves participate in losses, while space waves and leaky waves are related to radiation. Leaky waves are more directional than space waves. Therefore, the gain may be increased by combining leaky and radiation waves. The size of the antenna used in this device can thus be minimized. Changes in efficiency and directionality can be achieved by controlling bound wave and leaky mode limitations, respectively.

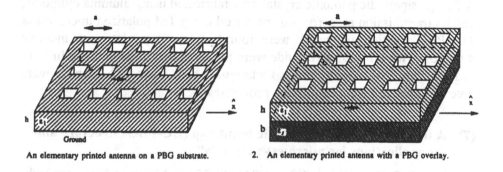

An elementary printed antenna on a PBG substrate.   2. An elementary printed antenna with a PBG overlay.

Figure 5.11(4-1) Photonic crystal (left) and that with a dielectric underneath (right).

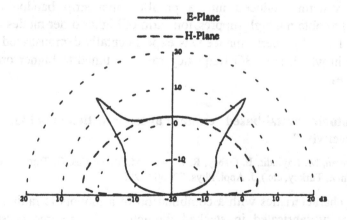

Figure 5.11(4-2) Calculated radiation patterns.

(5) "Infrared filters using metallic photonic band gap structures on flexible substrates"

S. Gupta, G. Tuttle, M. Sigalas, and K. Ho (Iowa State Univ., USA), Appl. Phys. Lett. **71**, 2412 (1997).

A metallic photonic crystal which has a PBG in the infrared range is discussed. As a high pass filter, it operates with a cut-off frequency of 3 THz and an attenuation of over 35 dB in the cut-off region.

(6) "Band gap and wave guiding effect in a quasiperiodic photonic crystal"

C. Jin, B. Cheng, B. Man, Z. Li, D. Zhang, S. Ban, and B. Sun (Chinese Academy of Sci., China), Appl. Phys. Lett. **75**, 1848 (1999).

A 2D quasiperiodic photonic crystal was fabricated using alumina cylinders, and its transmission property was measured using TM polarized microwaves. The PBG position and width were found to have no relation to incident direction. Two kinds of waveguide were prepared by removing three lines to form a linear waveguide and a rectangular bent waveguide, which were revealed to have high transmission properties.

(7) "A uniplanar compact photonic band-gap (UC-PBG) structure and its applications for microwave circuits"

F. R. Yang, K. P. Ma and T. Itoh (UCLA, USA), IEEE Trans. Microwave Theory and Tech., **42**, 66 (1999).

A PBG structure induced into a parallel microstrip bandpass filter is discussed to obtain a high suppression ratio of higher order modes and a low insertion loss. High performance was experimentally demonstrated in such a structure in which the PBG frequency range is tuned to higher order mode frequencies.

(8) "Photonic crystal-based resonant antenna with a very high directivity"

B. Temelkuran, M. Bayindir, E. Ozbay, R. Biswas, M. M. Sigalas, G. Tuttle, and K. M. Ho (Bilkent Univ., Turkey, etc.), J. Appl. Phys. **87**, 603 (2000).

Antenna characteristics with a combination of a monopole radiating source and a cavity fabricated in stacked dielectric 3D photonic crystals were investigated. The sharp directionality of this antenna was found to agree well with the theoretical predictions. The antenna has a narrow band determined by the cavity.

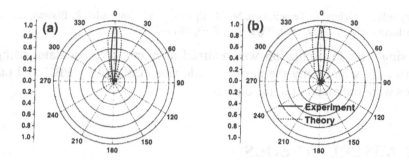

Figure 5.11(8) Radiation patterns showing a steep directionality.

(9) "Photonic-band-gap resonator gyrotron"

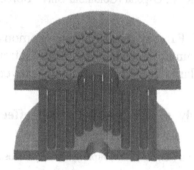

Fig. 5.11(9-1) Structure of gyrotron with 2D photonic crystal.

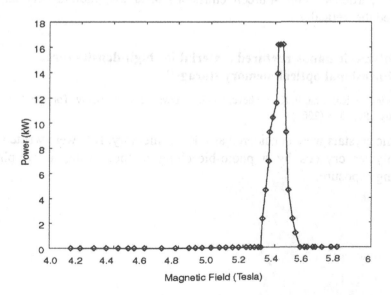

Fig. 5.11(9-2) Output power versus magnetic field intensity.

M. Bayindir, E. Ozbay, B. Temelkuran, M. M. Sigalas, C. M. Soukoulis, R. Biswas, and K. M. Ho (Bilkent Univ., Turkey, etc.), Phys. Rev. B **63**, 081107 (2001).

The single mode oscillation was realized at microwave oscillator using a gyrotron, in which only the desired mode was confined by PBG and other modes were radiated out to the outside.

5.12 MISCELLANEOUS

(1) "Photonic bandgaps with defects and the enhancement of Faraday rotation"

M. J. Steel, M. Levy, and R. M. Osgood (Columbia Univ., USA), J. Lightwave Technol. **18**, 1297 (2000).

The enhancement of Faraday rotation was demonstrated to occur in a photonic crystal containing defects. The introduction of two defects was shown to enable both large Faraday rotation and power output.

(2) "Theoretical study of the Smith-Purcell effect involving photonic crystals"

K. Ohtaka, and S. Yamaguchi (Chiba Univ., Japan), Opt. Spectrosc. **91**, 477 (2001).

The radiation of electromagnetic waves that accompany the movement of charged particles (Smith-Purcell emission) on a 2D photonic crystal was analyzed theoretically.

(3) "Polymeric nanostructured material for high-density three-dimensional optical memory storage"

B. J. Siwick, O. Kalinina, E. Kumacheva, and R. J. Dwayne miller (Univ. Toronto, Canada), J. Appl. Phys. **90**, 5328 (2001).

Photonic crystals were considered as a ROM memory. Bits were written on a 3D polymer crystals by a photo-bleaching method using a two-photon scanning exposure.

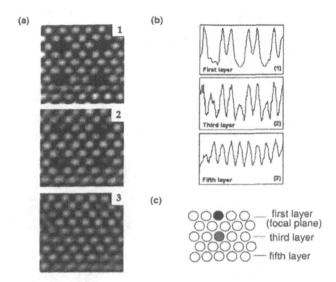

Fig. 5.12(3) (a) SEM image showing written bits by two-photon scanning in the first, third, and fifth layers. (b) Emission intensity profiles of the first, third, and fifth layers. (c) Cross-sectional image of a typical model of the samples.

Chapter 6

TECHNOLOGY ROADMAP ON PHOTONIC CRYSTALS

This chapter shows a roadmap table which summarizes the so-far obtained results in many research organizations and future prospects for each application. It includes each ultimate performance expected within next 10 years and many issues to be investigated in the near future. Some of them are very general or well known among experts. Or some of proposed devices are just substitutes of existing products. However, such a roadmap will help the advancement of the research field and the development in industries.

Note that owing to the rapid progress of the research, many of the listed items in the roadmap have been realized between 2 to 5 years earlier than anticipated.

(by Editors)

Year	– 2002	2004
Light Emitters		
LED	High efficiency demonstrated. Current injection demonstrated. Surface plasmon effect discussed.	Ultimate efficiency estimated. Total design, various materials. Commercialization starts.
Band edge laser	Lasing by current injection with $I_{th} < 100$ mA. 2D mode control in large LD. Polarization control demonstrated.	Structure optimized. Low threshold, high power.
2D defect laser	Lasing in single defect. Enhancement of Q-factor discussed Mode control discussed.	Current injection demonstrated. Integration with waveguide. Total discussion of performance.
3D defect laser	Perfect 3D PBG demonstrated. Defect level emission observed.	Lasing by photopumping. Spontaneous emission control.
Other lasers	Quasi-PC, PC VCSEL, microgear cylinder grating demonstrated.	Mode analysis and optimum design. Investigation for photonic integration
Fibers		
Holey- and PBG-type	Min. loss < 1 dB/km. Dispersion compensation discussed. Wide band single mode in large core. Large nonlinearity for supercontinuum, etc. Various doping and functions.	Total discussion on performance for various types. Partly commercialized in transmission line.
Photonic IC		
Waveguide type	2D PC waveguide demonstrated. Sharp bend, branch, coupler, etc. discussed. Propagation loss measured. Connection to fiber discussed. 3D waveguide and bend designed.	Loss reduction to 1dB/cm. Ultimate low loss discussed. Low loss in micro-elements demonstrated. 3D PC waveguide demonstrated.
Bulk type	Superprism type discussed.	Loss evaluation in superprism. Composite prism synthesized.
WDM components		
Resonant filter	Various add/drop waveguide filter discussed. Surface coupled-type defect filter demonstrated. High Q over 3000 demonstrated. Autocloning type demonstrated.	Synthesis of filter function demonstrated. High efficiency I/O to fiber developed.
Waveguide type filter	2D PC waveguide demonstrated.	Basic function demonstrated. Detailed dispersion discussed .
Superprism filter	DEMUX function demonstrated. Add/drop filter proposed. Optimum structure discussed.	High resolution demonstrated. High efficiency I/O developed.
Polarizer	Planar polarizer by autocloning developed. Partly commercialized.	Simple process for cost reduction. Polarization rotator discussed.

2007	2010 –
Max. efficiency of 30 – 50 % achieved. Simple production method developed. High speed operation discussed.	Efficiency > 50%, high speed ×10. Beam control, etc.
Stabilization of single mode. Diffraction limit spot demonstrated. High power > 100 mW.	High power > 1W. Perfect single mode.
Threshold < 10 μA demonstrated. Large array, athermal operation, etc.	Threshold < 1 μA. High density integration of >10000 devices.
Current injection lasing. Systematic 3D fabrication established. Progress similar to PC laser expected.	Ultra-low threshold <<1 μA. Ultimate performance in various aspects. Progress toward integrated light source.
Ultimate low loss achieved. Widely commercialized.	< 0.2dB/km. > 1W propagation. >100 mm core. UDWDM, soliton, etc.
Large tolerance for design, fabrication, etc. Photonic IC with 100 elements. Integration with active components. Partly commercialized.	Photonic IC with 10,000 components. Multifunctional platform.
Low loss and efficient I/O realized. Packaging discussed.	Total insertion loss < 1dB. Photonic IC for special targets.
Integration of > 100 channels. Large tolerance, trimming, athermal. Partly commercialized.	Add/drop filter for >2,000 channels
Low loss waveguide used. Size reduction with many I/O ports.	Port numbers >> 1,000 in 1 mm^2 with various functions.
Simple production developed. Bulk device partly commercialized. Add/drop device demonstrated.	Linewidth of < 0.4 nm in 1mm^2. Add/drop with < 3 dB insertion loss.
Polarization rotator demonstrated. Low cost circulator.	Widely commercialized with many functions.

Year	– 2002	2004
Light Control		
Switch	Use of nonlinearity discussed. Device configuration discussed.	Switching function demonstrated. Ultimate performance discussed. Matrix arrays discussed.
Wavelength tuner	Tuning by liquid crystal demonstrated. Tuning by carrier effect discussed.	Materials for large tunability discussed. Tunable laser and filter demonstrated.
Memory and delay		
Defect memory	Ultra-high Q cavity for ns memory discussed.	Ultimate Q calculated.
Delay line	Low group velocity at band-edge demonstrated. Pulse delay in 1D coupled cavity oberved.	Pulse delay demonstrated with dispersion management. Delay line by PC fiber investigated.
Packaging		
Connection	Reduction in reflection discussed. Spotsize converter by superprism discussed. Low loss connection with fiber demonstrated.	Systematic analysis carried out. Low loss < 0.3 dB demonstrated.
Dispersion compensator		
PC waveguide-type	Use of waveguide mode discussed. Coupled cavity waveguide (CCW)-type discussed.	Optimum band structure investigated. Design theory established. Dispersion compensator demonstrated.
PC bulk-type	Use of band structure discussed. Superprism compensator discussed.	Design and analysis.
Fiber-type	Refer to 'fibers'.	Refer to 'fibers', hereafter.
Quantum information and communication		
Quantum devices	Squeezing, single photon emission discussed.	PC single photon emitter demonstrated. PC quantum gate investigated.
Storage		
Focusing system	Focusing by negative refractive index discussed. Superlens for focusing suggested.	Large NA demonstrated
MO disk	1D magneto-optic PC for large Faraday rotation discussed and partly demonstrated.	Effects with 2D/3D PC discussed. Configuration of head and disk discussed.
Read/write head	Use of planar polarizer discussed.	Simplified process developed.
Light source	Refer to 'Light emitter', hereafter.	
SHG source	SHG by multiple photonic bands demonstrated. Easy phase matching demonstrated.	Quantitative evaluation of efficiency in 2D PC.
Deflection/scanning		
Prism type	Deflector by superprism discussed. High efficiency interfaces investigated.	Basic function demonstrated. Integration of emitter.

2007	2010 –
Energy consumption of fJ order realized. Photonic router discussed.	Large-scale switch arrays.
Optimization and stabilization.	Commercialized.
Q>>10,000 demonstrated. Memory and read demonstrated.	Memory operation >>ns realized.
Partly commercialized. Tunability demonstrated. Integration with other elements	Delay time >>ns achieved
Loss further reduced. Packaging developed.	Various connections with < 0.1 dB loss. Alignment free connection demonstrated.
High efficiency I/O. Bulk device partly commercialized. Low loss, athermal, tuning demonstrated.	Arbitrary dispersion compensation.
Fundamental function demonstrated. High efficiency I/O, tuning discussed.	Arbitrary dispersion compensation.
High efficiency, high speed realized. Quantum gate device demonstrated.	Basic devices in quantum computing and quantum communication systems.
Resolution limit spotsize demonstrated. Combination with nonlinearity.	Low cost PC lens installed on disk. > 100Gbit/inch^2 storage achieved.
Read out experiment.	×100 Faraday rotation for high density Storage.
Commercially available. Integration of R/W head.	Simple compact head by autocloning and related self-assembled technologies.
Demonstration of ideal efficiency.	SHG efficiency > 90% with low power. UV light by SHG.
30° deflection in compact device.	Implementation for printer, scanner, etc. Used as a spatial switch.

Year	− 2002	2004
Electromagnetic wave application		
Micro/millimeter wave	Metal/dielectric PC studied. Filter operation demonstrated. High performance antenna demonstrated.	Easy production developed. Antenna performance improved. Packaging method developed.
Light-microwave conversion	Conversion method discussed.	Tuning of metal PC with PD demonstrated. Converter with antenna designed.
THz devices	THz generator with 3D PC demonstrated.	High efficiency generation realized.
Others		
Illuminator	Enhancement of thermal emission spectrum by band-edge demonstrated.	Mechanism analyzed. Application to projector back light discussed.
Heat control	Control of heat discussed.	Design of PBG. PBG effect demonstrated.
Image processor	Negative refractive index for image formation, etc. discussed.	Negative index demonstrated. Anti-reflection investigated.

2007	2010 –
Commercially available. Size reduction, tunability, etc. Installation into mobile phone.	Functional antenna improved from present performance by 5 – 10 dB.
Fundamental function demonstrated. Low loss, high efficiency, miniaturization.	High efficiency direct conversion of lightwave to microwave.
THz sensor investigated.	High efficiency emission and detection achieved.
Ideal characteristics demonstrated.	Ecological bright illumination.
Application to heat insulator and heat pump demonstrated.	Control of heat, low cost.
3D TV, hologram, optical computing investigated.	Various applications discussed for new optics.

INDEX